LEAN *and* CLEAN
MANAGEMENT

Joseph Romm

LEAN *and* CLEAN MANAGEMENT

How to Boost Profits and Productivity by Reducing Pollution

JOSEPH J. ROMM

KODANSHA INTERNATIONAL
New York • Tokyo • London

The author currently serves as Special Assistant to the Deputy Secretary for the U.S. Department of Energy. This book was written prior to the commencement of his service to the Department. The views and opinions expressed in this book are not to be construed as necessarily reflecting those of the Department. Furthermore, the information contained in this book does not draw substantially on ideas or official data that are federal nonpublic information.

Kodansha America, Inc.
114 Fifth Avenue, New York, New York 10011, U.S.A.

Kodansha International Ltd.
17-14 Otowa 1-chome, Bunkyo-ku, Tokyo 112, Japan

Published in 1994 by Kodansha America, Inc.

Printed in the United States of America

94 95 96 97 98 99 7 6 5 4 3 2 1

Library of Congress Cataloging-in-Publication Data
Romm, Joseph J.
Lean and clean management : how to boost profits and productivity
by reducing pollution / Joseph J. Romm.
p.   cm.
Includes bibliographical references and index.
ISBN 1-56836-037-1
1. Industrial management—Environmental aspects. 2. Environmental
policy—Economic aspects. 3. Environmental protection. I. Title.
HD69.P6R66 1994
658.5′67—dc20   94-17886
CIP

Book design by Jessica Shatan
The text of this book was set in Bodoni Book
Composed by Haddon Craftsmen, Scranton, Pennsylvania
The jacket was printed by Phoenix Color Corporation,
Hagerstown, Maryland
Printed and bound by R. R. Donnelley and Sons,
Harrisonburg, Virginia

To Ethel Grodzins Romm and Al Romm,
the two finest managers I know

It is the little, pathetic attempts at Quality that kill.
—ROBERT M. PIRSIG
  Zen and the Art of Motorcycle Maintenance

Unless all sources of waste are detected and crushed, success will always be just a dream.
—TAIICHI OHNO
  Toyota Production System

Contents

Acknowledgments   xi

Preface   THE FASTEST IN THE WEST—AND THE GREENEST   xv

Introduction   LEAN AND CLEAN MANAGEMENT   3

1   THE AMERICAN ORIGINS OF LEAN AND CLEAN   16

2   SYSTEMS: THE CYCLE OF LIFE   31

3   LABOR AND LEAN AND CLEAN   50

4   A FOCUS ON PREVENTION: THE CASE OF COMPAQ
      COMPUTER   63

5   ENERGY EFFICIENCY   72

6   LEAN AND CLEAN DESIGN FOR OFFICE PRODUCTIVITY   90

7   BECOMING LEAN AND CLEAN: A SYSTEMS APPROACH   105

8   LEAN AND CLEAN DESIGN FOR FACTORY PRODUCTIVITY   130

9   THE FUTURE IS LEAN AND CLEAN   167

Conclusion   MANAGEMENT, THE ENVIRONMENT, AND JOBS   177

Appendix   YOU JUST DON'T UNDERSTAND:
             U.S. MISPERCEPTIONS OF JAPANESE SUCCESS   183

Notes   187

Bibliography   205

Index   209

Acknowledgments

This book is my attempt at a systematic management strategy that will enable a company to increase profits and productivity by reducing pollution. It was possible only because I have been able to work with and learn from practitioners in a variety of disciplines from the private, public, and nonprofit sectors.

I am most indebted to Rocky Mountain Institute, and especially Amory Lovins, for instilling in me the essential idea of this book—that businesses can make money reducing pollution. Amory is not merely a brilliant thinker but a brilliant *systems* thinker. I am grateful to John Barnett for giving me the freedom to pursue this project, to Jennifer McCulloch for helping me track down some of the case studies, and to Andrew Jones for his insight into systems thinking. I owe special thanks to Bill Browning for sharing his knowledge of green building design and for collaborating with me on the peer-reviewed analysis of the connection between energy-efficient office design and productivity. Many other people at the Institute were generous with their time and their ideas, including Rick Heede, Michael Kinsley, Alice Hubbard, and Clay Fong.

This book grew out of a report I did for the Environmental Protection Agency and the Small Business Administration, which was made possible by the support and effort of David Wann and David Leavitt.

I am indebted to all the managers, engineers, designers, and line employees who shared their experiences with me and convinced me both of the need to write a book and of the need to focus on management issues. I would particularly like to thank John Carter, Steve Cassens, Joseph Crowley, Doug DeVries, John Donoghue, Lawrence Friedman, Ron Gonsalves, Michael Jaeb, Harold Kelly, Tachi Kiuchi, Eng Lock Lee, Robert McLean, Ken Nelson, Ron Perkins, Paul Scanlon, Ken Teeters, and Karney Yakmalian.

Special thanks go to Lee Windheim, who helped me see that the connection between energy-efficient office design and productivity not only was real but could be documented and explained. This essential part of my book would not have existed had he not spent the time to help me track down several of the cases published here for the first time.

I am grateful to Art Rosenfeld and Steven Ternoey for their comments on these cases.

I would also like to thank Stanley Boliver of Productivity Press for helping me find a variety of books on the Japanese approach to raising productivity through systematic process redesign.

I am grateful to Tad Smith for sharing his expertise in management, business, and economics and his insight into what is and is not original in a management book. My ideas benefited greatly from discussion with a variety of people knowledgeable about business and environmental issues, including Braden Allenby, Don Aitken, Mirka della Cava, Ted Flanigan, and Hardin Tibbs.

If this book presents a coherent worldview of management, several people share the credit. Several years ago I was fortunate to hear Col. John Boyd deliver a brilliant series of lectures, "A Discourse on Winning and Losing," in which he integrated information from several fields. In a world of analysts, he is that rare breed—a synthesist. He directed me to the best thinking on lean, fast-cycle management. I am forever in his debt for his insisting that I read the work of Taiichi Ohno and Shigeo Shingo and learn what the Japanese thought they did, as opposed to what American management consultants thought the Japanese did.

My mother—a designer, a writer, a building manager, and a chief executive officer—has always shared her extraordinary wisdom with me. She improved the book immeasurably by applying her remarkable language and management skills to innumerable drafts. For thirty years my father was a newspaper editor, perhaps the definitive fast-cycle management job. Seeing him lead the production of an entirely new product every day, 365 days a year, was probably how I first realized that speed and quality could go hand in hand. I also cannot adequately thank him for focusing his world-class editorial skills on this book. I owe a special debt to Kate-Louise Gottfried for her support and understanding during the writing of this book.

I am very grateful to John Urda, my editor at Kodansha, whose insight improved the writing and the structure of the book. I would also like to thank my copyeditor, Susan M.S. Brown.

Finally, special thanks go to Peter Matson, my agent, who continues to make it look easy to find a good publisher for my writing.

THE FASTEST IN THE WEST—
AND THE GREENEST

In 1986 the mail sorters at the Reno, Nevada, main post office became the most productive—and the most error free—of all the sorters in the western half of the United States. Was this remarkable result made possible by the introduction of a new quality-oriented management initiative? Did some of the operators of the mail-sorting machines receive special training? Were they part of an experiment designed to boost productivity?

Not at all. In fact, the director of mail processing, Robert McLean, denies any personal responsibility for the improvement. McLean, now postmaster for Carson City, says, "We had the same people, the same supervisor, and I don't believe I was doing any motivational work." Yet he says that the data on the productivity and quality increase were solid: "It was irrefutable."[1]

What happened? It had begun a few years earlier when the Reno Post Office was selected to receive a renovation that would make it a "minimum energy user." An architectural firm, Leo J. Daly, was hired to design changes aimed at doing everything necessary to reduce energy use.

The post office was a modern warehouse with high ceilings and a coal-black floor. It was quite noisy in the areas where the two sorting machines were run. The chief architect, Lee Windheim, proposed a lowered ceiling and improved lighting (plus other energy-saving measures). With the new ceiling, the room would be easier both to heat and cool and, additionally, have better acoustics. The ceiling would be sloped to enhance the new indirect lighting, which would replace the old, harsh direct downlighting with softer, more efficient, and longer lasting bulbs. The energy-efficient room would be far more pleasant to work in.

Before starting the complete renovation, which would cost about $300,000, Windheim did a mock-up of the lighting and new ceiling. The idea was to let it run for a few months to see how it worked and how people liked it. The mock-up was installed over only one of the two sorting machines. The graph on this page shows the number of pieces of mail sorted per hour in the twenty-four weeks before the change, and for more than a year after the change.

In the twenty weeks following the renovation, productivity shot up more than 8 percent, while the machine in the area with the old ceiling and lighting showed no change in productivity. Ten months later, as the graph shows, productivity had stabilized at an increase of about 6 percent. A postal worker operating the machine was now sorting about 1060 pieces of mail in the time it used to take to sort 1000.

The sorter is grueling to use. Once a second it drops a letter in front of the operator, who must punch in the correct zip code before

PRODUCTIVITY RISE FROM ENERGY-EFFICIENT DESIGN

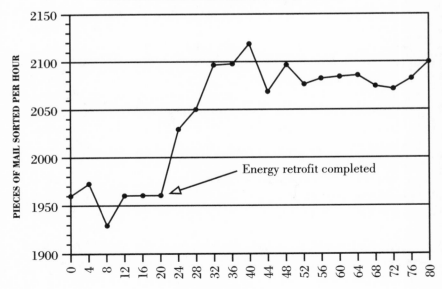

the next letter appears. If the operator keys in a zip code that doesn't exist, or no zip code at all, the letter will immediately be sent through the machine for repunching. If the wrong zip code is keyed in, the letter will be sent to the wrong bin, and it will take even longer to track down the mistake. The job is so intense that an operator can work only a maximum of thirty minutes on the machine at one time.[2]

After the mock-up, the rate of sorting errors by machine operators dropped to 0.1 percent—only one mistake in every 1000 letters—the lowest error rate in the western region. As McLean tells it, "No one could poke holes in the story." The data were "solid enough to get $300,000 to do the whole building." After the renovation, "people used to hang out there after work. It wasn't just the lighting, it was the whole impact on the work environment. But the lighting was the main thing."

The energy savings projected for the whole building came to about $22,400 a year. The new ceiling would bring an additional savings of $30,000 a year by reducing the recurring maintenance cost of repainting the underside of the exposed roof structure. Combined, the energy and maintenance savings came to about $50,000 a year: a six-year payback. That is, the energy and maintenance savings would cover the cost of the improvements in six years.

The productivity gains, however, were worth $400,000 to $500,000 a year. In other words, *the productivity gains alone would pay for the entire renovation in less than a year.* The annual savings in energy use and maintenance and the reduced air pollution were a bonus. Working in a quieter and more naturally lit area, postal employees did their jobs better and faster. Designing for the end user, in this case the postal worker, always costs the least and invariably increases productivity.

The Reno Post Office became not only the most energy-efficient, environmentally benign post office in the western region, as intended, but also the most productive and error free.

It became lean and clean.

We know from work done at Western Electric in the 1930s that contrived experiments to monitor the effect a workplace change has on productivity can be complicated by the special conditions of the experiment, particularly the interaction between the worker and the experimenter (see Chapter 6). But at the Reno Post Office no one conducted a special experiment intended to raise productivity, nor was there unusual interaction between workers and supervisors. The changes were designed solely to reduce energy use and air pollution. Productivity had always been monitored, and the improvements were unexpected.

The story of the Reno Post Office has never been told before. Not long afterward, the post office was reorganized, and individuals moved to other jobs or retired. The word never got out.

This is but one of many case studies—documented for the first time in this book—showing that designing buildings and offices to reduce energy use and pollution boosts worker productivity. We will see how a major defense contractor, an insurance company, and a bank all used "green" building design to increase worker productivity dramatically and reduce absenteeism. We will also see how manufacturing companies such as Compaq Computer have used energy efficiency, pollution prevention, and environmental design to raise productivity and profits.

Lean and clean is far more than a management philosophy. It is the way everyone in every company—and every government—must learn to act and think in order to thrive in the next decade and the next century. Workers will need to become expert in reducing nonlabor costs—the essence of lean and clean—to help preserve their jobs and increase wages. Engineers and architects who want a competitive advantage will need to master lean and clean. Universities will need to train their students for the rapidly growing field of pollution prevention and lean and clean management. Policy makers in national and local government will want to use regulatory and technology policy to ensure that their businesses become supremely competitive through clean production and that their workers capture the rapidly growing world market for clean technologies.

A variety of books, articles, and studies have made clear the tremendous cost savings from reduction of resource use and pollution. This book is the first to document that pollution prevention by service companies as well as manufacturers doesn't just save money but it invariably, if unexpectedly, raises productivity, which raises savings even more dramatically than anticipated. This book will show not only *what* measures boost productivity but *why*.

LEAN *and* CLEAN MANAGEMENT

LEAN AND CLEAN MANAGEMENT

If you think about it, pollution is fundamentally a manifestation of economic waste. It involves incompletely using a resource, throwing a resource away, or burning something. The opportunity to lower cost by eliminating pollution, then, seems anything but rare.
—MICHAEL PORTER, Harvard Business School[1]

E very company can increase its profits and productivity dramatically by reducing pollution. Every company. The *technological* problems are not the biggest barriers, the *institutional* problems are—but all barriers can be overcome through lean and clean management.

The companies that have applied this approach in even an elementary way have reduced their resource use and waste disposal costs by 50 percent or more with rapid payback. Those changes not only prevented pollution but often dramatically improved productivity, reducing the payback to under one year. If every business in the country matched the gains of the best companies, we would increase our international competitiveness, accelerate productivity growth, generate high-wage jobs, and reduce our air pollution, water pollution, and solid waste by 50 percent or more.

The idea behind lean and clean management is simply stated: to reduce waste, particularly wasted resources. Most companies produce very little of either the energy or the material they use. Therefore, to the extent that a company's designers, engineers, managers, and line workers figure out how to reduce that energy and material use (and hence prevent pollution), they will be keeping money that once left the company, allowing both profits and wages to rise. Lean and clean companies reduce disposal costs, avoid fines, and minimize negative publicity. Such companies boost productivity by eliminating inefficiency, streamlining production, improving the workplace environment, and increasing wages. Lean and clean companies become supremely competitive.

The most compelling reason to embrace pollution prevention is that it pays. A 1992 cross-industry survey of seventy-five case studies of pollution prevention found an average payback for industrial investments in waste reduction of only 1.58 years: an annual return on investment of 63 percent. The Louisiana Division of Dow Chemical has audited 575 energy- and waste-reducing projects since 1982. The average annual return on investment for those projects came to an astonishing 204 percent. Annual savings exceed $110 million.[2]

Consider the results of a study of eighty-four companies comparing those who dealt with pollution proactively through recycling, energy efficiency, and waste minimization with those who had consistently poor environmental records, including repeated regulatory violations and/or major actions. The environmentally proactive companies had 4 percent higher return on investment, 9 percent higher sales growth, and almost 17 percent higher operating income growth.[3]

The number of win-win examples of proactive companies is ballooning:

- **Harrah's Hotel and Casino** in Las Vegas was able to reduce its energy and water costs for cleaning sheets by $70,000 a year.

- **Lockheed** built an engineering development and design facility that saved $300,000 to $400,000 a year on energy bills—and productivity rose 15 percent.

- **Republic Engineered Steels** saves nearly $3.6 million a year by recycling its steel more efficiently.

- **Boeing** reduced its lighting electricity use by up to 90 percent with a two-year payback—a 53 percent return on investment. The new, higher quality lighting cuts down glare and helps workers reduce defects.

• The ten-story **Comstock** building went from design through construction in eighteen months, yet came in $500,000 under budget and has half the operating cost of a typical Pittsburgh building.

• The **UKettle** boils water faster and with less energy than stove-top models and microwaves. Unlike most products, the UKettle was designed with parts that are easy to separate and recycle. Yet it was brought to market in eight months, half the time usually needed for small appliances.

• By 1992, **Baxter Healthcare** was recycling 99.9 percent of its plastic scrap. Recycling has saved the company $9 million over the last decade.

How were these feats achieved?

• **Harrah's** simply asked its customers—its guests—whether they wanted their sheets changed every day. The vast majority said no.

• **Lockheed's** architects designed a building that improved the workplace environment, making greater use of daylight and increasing opportunities for worker interaction.

• **Republic Engineered Steels** got the idea for improved recycling from an employee suggestion system.

• **Boeing** worked with the Environmental Protection Agency to cut energy use and took a systematic approach to lighting redesign.

• The **Comstock** building had a design team that included representatives from all architectural and engineering disciplines who worked together from the onset of the project.

• The **UKettle** design team included an applications engineer, mechanical engineer, industrial designer, and tooling specialist. Design for ease of disassembly is the flip side of design for assembly.

• **Baxter** focused its quality circles on waste reduction, recognized employees who contributed to that goal, and committed to constant, incremental improvement in scrap reduction.

Do these practices of energy efficiency and clean production sound familiar? Asking customers what they want, following employee suggestions, making use of cross-functional teams, designing for assembly, improving constantly, and taking a systems approach—these are the same practices used by companies that have successfully embraced popular new approaches such as Total Quality Management, reengineering, and lean production. Success in each area requires a systematic approach, an examination of the entire production or management process to eliminate waste. The reason lean production is compatible with clean production is that both have the same goal: systematically reducing waste.

In the case of lean production and Total Quality, the waste is *wasted time,* and the measures of inefficiency are high inventories, defects, and customer complaints. In clean production, the measure of inefficiency is pollution—air pollution, water pollution, and solid waste. If a company has successfully improved quality and reduced wasted time, lean and clean management is the next step in the ongoing process of increasing profits and productivity.

But most companies fail to improve quality or reduce wasted time in either a significant or an enduring fashion. Many don't try, under the mistaken belief that it is not worth the trouble. Others try but fail, principally, I believe, because their management and their consultants don't really understand how and why they must change their manufacturing or service delivery processes in a systematic fashion. This book will explain what systematic process improvement is, why it is the only way to remain competitive in the long run, and how to apply it to reducing both wasted time and wasted resources.

The simplest resource to conserve is energy. Energy use is a dynamic system: The heat produced by excess lighting is one of the biggest contributors to air-conditioning costs. Integrating lighting improvements with other changes, including better insulation and window coatings, can cut heating and cooling bills by 50 percent or more, not only earning a rapid payback and improving comfort and working conditions but also potentially increasing productivity and reducing absenteeism.

Pollution prevention of any kind requires a systemic approach. Many companies thought that doing without chlorofluorocarbons (CFCs) would be exceedingly expensive since these substances had been an integral part of their production processes for years. But after examining their processes from beginning to end, they found that systemic process redesign would actually allow them to eliminate CFCs while *lowering* overall costs.

Lean production and clean production can be integrated into an overall approach, called lean and clean management or total quality environmental management. The same cross-functional teams needed to reduce product cycle time are needed to reduce pollution. For example, lean production requires designing products that are easy to *put together,* while clean production requires designing products that are easy to *take apart* (to promote reuse and recycling). These two goals can be achieved jointly with lean and clean design that emphasizes simplicity and modularity.

Lean and clean management is the most comprehensive approach to minimizing all types of company waste. It is, therefore, likely to become the dominant management and production system of the twenty-first century.

If America does not become the leader in lean and clean management, we run the risk of permanently undermining our economic health, for two reasons. First, if other countries lead the way in the development of high-productivity, low-cost clean processes and products, American businesses will suffer competitively, just as they did for being so late in the lean production revolution. Second, as developing nations realize they do not have to develop as we did, they will leapfrog directly into resource-efficient growth, and they will buy products and processes from foreign firms already practicing lean and clean (see Chapter 9). Indeed, if our businesses are slow to catch on to the clean revolution, the way they were with the lean revolution, *we* will be importing those products and exporting jobs. That unfortunate trend started two decades ago, when we began importing fuel-efficient cars from Japan, and it continues today—we import about 70 percent of our air pollution control equipment from Germany.[4]

American business has already downsized twice in the past decade and a half because it was so slow to adapt to change. In the late 1970s and early 1980s, we lost a big chunk of our manufacturing industry. Again in the late 1980s and early 1990s, we lost a big chunk of our high-wage white-collar service jobs, together with another chunk of high-wage manufacturing jobs. Yet the problems facing this country continue to be severe: low productivity growth and high resource use. The great risk is that if U.S. businesses don't change, they will downsize yet again, at the turn of this century, destroying the American dream for another generation or more.

PREVENTING JOB LOSS

Lean and clean management saves jobs by replacing capital expenditures and resource consumption with human ingenuity and labor. Employees will figure out how to reduce the use of expensive resources and stop the hemorrhaging of money that accompanies waste. Brainpower replaces resource use and prevents pollution in this new industrial revolution. Lean and clean saves jobs by reducing the *nonlabor* costs of doing business.

Downsizing is the rage in corporate America. Yet this "no pain, no gain" approach has three fatal flaws. First, simply downsizing does not address a company's systemic problems, making another downsizing in the next recession likely: *Flawed companies that downsize simply become smaller flawed companies.*

Second, downsizing removes a tremendous storehouse of brainpower and experience that could have been used to address systemic problems, cut costs, and improve productivity. Third, when employees see that the company considers them expendable, morale plummets. As Labor Secretary Robert Reich puts it, "Companies are cutting their hearts out." Not surprisingly, studies have found that downsizing often fails to create enduring increases in earnings, productivity, customer service, or financial performance.[5]

A company seeking to become leaner may well have to find new work for the legion of middle managers and indirect workers who are not adding value to the product. With lean and clean management, however, it may be possible to minimize or avoid the layoffs that Total Quality Management might otherwise have required. Before firing anyone, consider these options:

SAVE JOBS WITH LEAN AND CLEAN

• *Use workers' knowledge to cut resource use and costs.* Republic Engineered Steels kept layoffs to a bare minimum during the recent slowdown in steel by putting in place an employee suggestion system. The resulting suggestions, many aimed at recycling and pollution prevention, have already reduced annual costs $45 million and will eventually cut another $20 million or more.

• *Improve process efficiency.* A Sealtest ice cream plant on the verge of being shut down put in place a utility-financed modernization that reduced energy costs from 7.5 cents to 5.5 cents per gallon of ice cream and improved manufacturing productivity 10 percent. That competitive edge saved two hundred jobs and the plant is now adding new manufacturing workers.

• *Reduce energy use everywhere.* A comprehensive efficiency program saved the Southwire Corporation $40 million in energy costs from 1981 to 1988. During a rough financial period, the savings were almost equal to all of the company's profits and may well have saved the company: a total of 4000 jobs at ten plants.

Shrinking companies make it difficult for their employees to help them improve. Growing companies make clear to workers that they are the firm's most important asset—more important than energy and resource use or even capital equipment. Letting employees use their talents to save their jobs is a surefire way to raise productivity. Chapter 3 discusses how to do this.

Everyone in the company benefits from a cleaner environment. Thus, preventing pollution appeals to employees' finest instincts, making them feel better about the company, boosting morale, and improving productivity further. The central point, however, is *not* that businesses should invest in prevention to help the environment. While this is an important goal, it is not the primary goal of most companies. The main goal of most companies is finding investments that increase profits and productivity. The central point of this book is:

> **Pollution prevention has a much higher rate of return than most investments your company is now making.**

Further, there are three reasons why the high-yield investments discussed in this book beat any other investments with comparably high returns:

1. *No or Low Present Risk:* Lean and clean techniques and technologies earn high rates of return. The investments are AAA-rated because they have already been used successfully by a variety of companies. Many require little or no capital.

2. *Reduced Future Risk:* Companies are protected from a rise in energy taxes or resource prices and shielded from environmental regulations that grow stricter and more sweeping every day. Lean and clean reduces worker exposure to unhealthy working conditions. And, with greener products and operations, the company's public image is improved.

3. *Increased Productivity:* Environmental design of offices and buildings boosts worker productivity and cuts absenteeism. Redesigning the entire manufacturing process and service delivery to reduce waste and pollution can make productivity soar.

WHY CHANGE HAS BEEN SLOW

Since systematic process redesign to prevent pollution can raise profits and productivity while saving jobs and helping the environment, this logical question follows: If the advantages are so tremendous, why isn't everyone investing in energy efficiency, pollution prevention, and environmental design?

A lot of companies are. In addition to the previous short list, AT&T, Compaq, 3M, Dow, Du Pont, Martin Marietta, and Xerox, to name but a few, are applying lean and clean techniques. Yet even for most of them energy and resource efficiency is relatively new. Until recently there have been too few results to make a strong case for lean and clean. Few companies are changing systematically, so most published results do not match what the best companies have done.

In addition, people remember the gasoline lines, lowered thermostats, and sweaters of the 1970s and still think that saving energy requires sacrifice ("freezing in the dark"). One 1988 survey of small businesses found that "energy efficiency was thought to require turning down heat or turning off lights, and these were not considered acceptable options, because a cold, underlit store would discourage customers."[6] They are confusing energy efficiency and energy conservation. Ending that confusion is a key goal of Chapters 5 and 6.

The most important benefit of lean and clean—the boost in worker productivity—is not widely known. Chapter 6 presents perhaps the only comprehensive attempt to document this benefit, showing clear evidence that the few companies that have taken a systematic approach to green office and building design have boosted productivity 7 percent to 15 percent while cutting absenteeism 15 percent to 25 percent. *The profits created by the rise in worker productivity can exceed the energy savings by a factor of ten.* Chapter 8 will show how factory managers and workers, by preventing pollution, can increase productivity and quality and slash life-cycle costs.

When one's worldview is flawed, what is obvious is not always easily seen or done. It took American companies two decades to realize that giving their customers quality products and outstanding service will increase profits. Obvious, isn't it? Yet some of our largest companies have still not caught on. A 1991 survey of top American companies found that only 22 percent always or almost always use customers' expectations to help suggest new products and services (compared with 58 percent for Japanese companies)—even though this is a cornerstone of Total Quality.[7] Quality is not simply reducing defects; it is giving customers what they want. A holistic or systems approach is required.

Why should companies that were slow to understand the benefits of a commitment to quality and customer service be any quicker to understand the benefits of reducing pollution? After all, whereas quality may have the benefit of *seeming* obvious, resource efficiency does not, at least in America.

The United States became a major industrial power at the turn of the twentieth century, a time when it was the world's leading producer of copper, phosphate, coal, molybdenum, zinc, iron ore, lead, silver, salt, tungsten, petroleum, and natural gas; and number two in bauxite and gold. One study of the origins of American industrial success concluded that "the most distinctive characteristic of U.S. manufacturing exports was intensity in nonreproducible natural resources; furthermore, this relative intensity was increasing between 1880 and 1920." By the late 1920s iron and steel products, machinery, automobiles and parts, and petroleum products accounted for more than half of all American manufacturing exports. Automobiles contained roughly half their value in iron and steel, nonferrous metals, and other fabricated metal products.[8]

The country's prosperity was thus built on a foundation of natural resource abundance. Even as late as the 1950s, America was still the world's largest producer of oil, extracting twice as much as the Middle Eastern and North African states combined. It is not surprising, then, that American companies are "resource fat": They require twice as much energy to produce a dollar of goods as companies in other industrialized countries, and they produce roughly five times the waste per dollar of goods sold as Japanese companies and more than twice that of German firms.[9]

A heavy reliance on energy and resources—and the resulting pollution—is built into the basic paradigm of most American companies. For decades most U.S. companies dealt with defects in quality *after* they had been made. The focus was not on preventing defects; rather, it was on inspection and rework: an expensive, inadequate "end-of-pipe" approach to quality. Similarly, our paradigm of waste and resource abundance led most companies in the 1970s and 1980s to focus on dealing with pollution after it had been created—with landfills, treatment, incineration, and the like—the same expensive, inadequate end-of-pipe approach applied to waste.

This approach has created many institutional barriers that limit lean and clean solutions. The Congressional Office of Technology Assessment noted many of these barriers in its comprehensive 1994 report *Industry, Technology, and the Environment*:

> Many firms are unaware of pollution prevention opportunities or their relative merits over end-of-pipe solutions. . . . Responsibility for finding pollution prevention solutions may not rest with those

most capable of doing so [line workers, engineers]. . . . Most environmental managers have been trained in end-of-pipe practices and thus may overlook opportunities for prevention. . . . Most environmental consulting focuses on end-of-pipe treatment, while most environmental equipment vendors sell end-of-pipe equipment. . . . Capital accounting practices and capital availability may limit the adoption of even profitable prevention projects.[10]

Many if not most companies view environmental projects solely as a means to satisfy a certain regulatory requirement and fail to perform even simple financial analyses on alternative prevention projects. Regulators, too, have traditionally emphasized end-of-pipe controls, thereby creating de facto standards, as many books and studies have noted. This book focuses on the institutional changes and the paradigm shift that businesses need to make to increase profits and productivity no matter how quickly the regulators change.

CHANGE

Change is never easy.

Changing a flawed worldview is especially difficult. As science historian Thomas Kuhn has explained, even as rational a group as the scientific community is reluctant to embrace change, clinging to scientific theories and paradigms long after they cease to make sense.

All managers have a worldview, conscious or otherwise, that guides their thinking. These days, a manager's worldview is likely to be the traditional style—hierarchical, centralized, and compartmentalized—which is increasingly ineffective in today's marketplace. Many managers know this, yet, without a compelling alternative, change is nearly impossible. In a race you can't beat a horse with no horse—even when your horse is dying.

Kuhn argued in his classic book, *The Structure of Scientific Revolutions*, that "once it has achieved the status of a paradigm, a scientific theory is declared invalid only if an alternative theory is available to take its place."[11] Kuhn's theory provides a key reason why so many good management ideas are not working—and gives us a way to solve this problem. Consider the quintessential case of Total

Quality Management (TQM), which is failing to take hold in most companies.

Total Quality Management is one of a long line of great ideas that sweep through American business, especially in bad times, promising to cure all ills. But it fails to take hold in most companies in large part because it is presented not as a coherent management philosophy or paradigm but rather as a disconnected cookbook of practices. Total Quality Management seminars and books often provide lists of recipes, without menus or an understanding of nutrition. Many consultants don't understand TQM. A company can improve the quality of a "lemon" without paying any attention to marketing, research, design, sales, or even its customers. But changing what is wrong about the lemon is not the same as—and may be a long way from—delivering the *right* product or service.

Total Quality Management has been too narrowly named, wrongly stressing a single static feature, quality. For TQM to be a successful management philosophy, it must be a time-based, systems-oriented philosophy, a life-cycle philosophy. It must be about the entire production process and not merely about tinkering with individual operations. It requires addressing the root causes of problems through prevention rather than merely fixing symptoms. It needs continuous feedback from all participants. The pioneer of the TQM movement, W. Edwards Deming, made clear that "the job of management in education, industry, and government should be the optimization of a system."[12] Thus the *Total* in Total Quality Management.

One cannot expect to achieve systemic change throughout a company without a theory of change. Thomas Kuhn articulated one of the most powerful theories of change: People will not abandon their current worldview, no matter how poorly it is performing, until they have a new one. This book recasts TQM and pollution prevention as a systems approach, a coherent management paradigm or philosophy powerful enough to cause the abandonment of the old worldview. Most business books have been written for managers alone, since managers have exerted the most control over what happens in business. This book, too, is aimed at managers. But I am writing here to machinists, engineers, designers, salespeople, maintenance people—all employees—because success in lean and clean management requires everyone's attention and understanding.

Before examining the systems approach to becoming lean and clean, however, it is worthwhile to look at the rarely recognized American origins of lean production and clean production.

1

THE AMERICAN ORIGINS
OF LEAN AND CLEAN

It is not possible to repeat too often that waste is not something which comes after the fact. Restoring an ill body to health is an achievement, but preventing illness is a much higher achievement. Picking up and reclaiming the scrap left over after production is a public service, but planning so that there will be no scrap is a higher public service.

Time . . . is a method of saving and serving which ranks with the application of power and the division of labor.

These chapter epigraphs are not the recent words of an EPA waste expert or a Japanese management guru. We think of lean production as a Japanese invention of the 1960s and 1970s, and clean production as an even more recent invention, but the above words were written in 1926 by Henry Ford. The Ford Motor Company, as we shall see, is where the Japanese learned so much of what we are now trying to learn from them.[1]

"Don't reinvent the wheel" is a basic principle as old as the biblical book of Ecclesiastes: "There is nothing new under the sun. Take anything of which it may be said, 'Look now, this is new.' Already, long before our time, it existed. Only no memory remains of earlier times."[2] Let us examine history so that we waste no time trying to reinvent what already exists but rather can build on what has gone before. This chapter shows how Henry Ford precisely described and practiced lean and clean management and how Japanese management and production efforts took his (and other Americans') basic ideas and improved upon them.

Although Japanese companies are well known for their ability to take an American idea—for example, the VCR—and build it into a successful product, they are capable of much more. The Japanese have consciously transformed many American ideas about quality into a systems-oriented approach focused on time and aimed at improving the entire production process. By understanding how Japanese fast-cycle manufacturing increases productivity, we can learn why clean production also increases productivity.

FORD'S OBSESSION WITH WASTED RESOURCES AND TIME

Henry Ford fathered the U.S. industrial revolution. In many respects the American century began with his introduction of the Model T in 1908, coupled with the discovery of huge pools of oil in Oklahoma and Texas. The car itself was durable and easy to repair, even by the typical owner. Ford's manufacturing breakthroughs made the Model T affordable to the average worker. In 1914 Ford had 48 percent of the U.S. auto market and was able to make almost as many cars as the other 300 U.S. car companies combined, using only one-fifth of their total labor.[3]

Ford's Model T came at a time when America was the leading producer of the key resources needed for an automobile revolution: coal, iron ore, and petroleum. Yet while Ford led the nation to greater wealth by exploiting our seemingly unlimited natural resources, he was obsessed with reducing wasted resources. In his 1926 book *Today and Tomorrow*, in a chapter titled "Saving the Timber," he wrote,

> We treat each tree as wood until nothing remains as a chemical compound to be broken down into other chemical compounds which we can use in our businesses. . . . Why should not crating be done with the smallest instead of the largest amount of lumber? . . . Why should a crate or a packing box once used be considered only as so much waste to be smashed and burned?
>
> We are cutting farther into wood by using wherever possible burlap bags and cardboard boxes—the latter made from waste and in our own paper mill. . . . We have a positive rule in every factory and branch that each crate and box must be opened carefully without

breaking the wood. . . . All scrap wood eventually gets back to the wood salvage department.[4]

These techniques—using every bit of a raw material, minimizing packaging, reusing packaging, replacing raw materials with reusable or recycled products, and recycling as much waste as possible—would not become standard waste minimization practices for sixty years. Terms such as *clean technology* and *clean production* would not come into widespread usage until the 1980s, after more than a decade of environmental regulations helped reveal the true cost of pollution and hazardous waste.

With his techniques Ford was able to reduce his use of wood by two-thirds even as he doubled production. Ford distilled all the wood waste, including sawdust, shavings, chips, and bark. He recounted, "Each ton of waste wood yields 135 pounds of acetate of lime; 61 gallons of 82 percent methyl alcohol; 610 pounds of charcoal; 15 gallons of tar, heavy oil, light oils, and creosote; and 600 cubic feet of fuel gas."[5]

Ford understood the hierarchy of clean production. Avoiding waste is most desirable, and reusing waste—which finds value in the waste and avoids disposal costs—is next best. "Picking up and reclaiming" scrap is good; "planning so that there will be no scrap" is better.

He was equally obsessed with reducing wasted metal and other resources.[6] In a chapter titled "Learning from Waste," Ford wrote, "Our studies and investigations up to date have resulted in the saving of 80,000,000 pounds of steel a year that formerly went into scrap and had to be reworked with the expenditure of labour." *No change was too small if it reduced waste.* Ford saved nearly 300,000 pounds of steel a year by cutting the fan-drive pulley out of the scrap from the hand door stock, rather than out of new stock. He originally made eighteen pieces of one part from a metal bar 143 inches long but was able to save "more than two inches per bar" when he found he could get the same number of pieces out of a bar 140%2 inches long.

Ford understood the importance of squeezing every last drop of useful work out of resources. Every ton of iron leaving the blast furnace came with over five tons of gas: 200,000 cubic feet. The hot gas was "cleaned and filtered to remove blast furnace dust and part of it used to pre-heat the blast. The balance is piped to the power house, where it forms the principal fuel." The blast furnace dust was nearly half pure iron—once regarded as waste, it too was reclaimed.

Although Ford pioneered the reduction of wasted resources, he may be best known for his efforts to reduce wasted time. His two great achievements—interchangeable parts and the moving assembly line—were tremendous time-savers. In 1908 Ford workers had to gather the needed parts, file them down and adjust them so they would fit, then put them all together. Each assembler would work on a large part of the car, perhaps spending all day on the mechanical system. The task cycle for a typical assembler—the time worked before repeating the same operation—was 514 minutes, or 8½ hours. Ford made the process more efficient by delivering parts to each workstation, minutely dividing the labor into individual tasks, and making the parts perfectly interchangeable (so they didn't need adjustment each time). Workers became far more efficient at their more specialized tasks. Cycle time plummeted from 514.0 to 2.3 minutes.[7]

Ford's greatest leap came in 1913: the moving assembly line. No longer would workers move from assembly stand to assembly stand. This cut cycle time still further—from 2.30 to 1.19 minutes. In under a year Ford reduced the effort needed to build major components by 60 to 80 percent. The time needed to assemble the major components into a complete vehicle dropped nearly 90 percent.

Ford's obsession with reducing wasted time went far beyond interchangeable parts and a moving assembly line. He wrote, "Time waste differs from material waste in that there can be no salvage." It is "the easiest of all wastes, and the hardest to correct" because "wasted time does not litter the floor like wasted material." At Ford, he said, "we think of time as human energy. If we buy more material than we need for production, then we are storing human energy— and probably depreciating its value."[8]

No price was too high to pay for saving time: "We have spent many millions of dollars just to save a few hours' time here and there." Ford's approach was systematic: Improving the cycle time for the entire production process was as important as improving the cycle time for individual workers.

FORD'S TIME-BASED SYSTEM

Our system . . . consists of planning the methods of doing the work as well as the work. . . . Our aim is always to arrange the material and the machinery and to simplify the operation so that practically no orders are necessary. . . . Our finished inventory is all in transit. So is most of our raw material inventory. . . .

Our production cycle is about eighty-one hours from the mine to the finished machine in the freight car, or three days and nine hours instead of the fourteen days which we used to think was record breaking. . . .

Let us say one of our ore boats docks at Fordson at 8:00 A.M. on Monday. . . . By noon Tuesday, the ore has been reduced to iron, mixed with other iron in the foundry cupolas, and cast. By 3 o'clock in the afternoon the motor has been finished and tested and started off in a freight car to a branch for assembly into a finished car. Say that it reaches the branch plant so that it may be put into the assembly line at 8 o'clock Wednesday morning. By noon the car will be on the road in the possession of its owner.

The man who wrote, "Having a stock of raw materials or finished goods in excess of requirements is waste," is rightly called the father of just-in-time. Taiichi Ohno, cofounder of the Toyota production system, once said that "if [the young] Henry Ford were alive today, I am positive that he would have done what we did."[9]

I would go further.

FORD ON LEAN AND CLEAN

If the young Henry Ford were alive today, he would no doubt be practicing and preaching lean and clean management. His writings make clear he understood that a focus on the end result demands such an integrated approach: "All men do not see the wisdom of fitting means to ends, of conserving material (which is sacred as the result of others' labours), of saving that most precious commodity— time; they must be taught." A contemporary British historian, J. A.

Spender, wrote of the production systems at Ford's remarkable River Rouge industrial complex, "If absolute completeness and perfect adaptation of means to end justify the word, they are in their own way works of art."[10]

In 1926 Ford wrote what might be called the credo of lean and clean:

> **You must get the most out of the power, out of the material, and out of the time.**[11]

Above and beyond his philosophy of eliminating wasted time and resources, Ford was committed to constant improvement. His most important lesson may well be never to rest on one's successes: "If we reach a stage in production which seems remarkable as compared with what has gone before, then that is just a stage of production and nothing more." The only lesson to be drawn from successful change is that more change is possible: "We know from the changes that have already been brought about that far greater changes are to come, and that therefore we are not performing a single operation as well as it ought to be performed."[12]

Ford was not, however, the perfect manager. His specialization of work turned what had been skilled work for artisans into highly repetitive and unsatisfying labor done at an ever-increasing pace. He understood processes but not people. His workers called him "the speed-up king." In the year Ford mechanized the assembly line, 1913, turnover in his highly paid labor force hit 380 percent! Soon, in order to keep a hundred men working, Ford was hiring almost a thousand.[13] Top wages were needed to attract men, but were never enough to keep them.

A key reason Ford was able to focus so much effort on improving the efficiency of the production process was that he built only one product—the Model T. It remained amazingly popular for many years, but he rigidly stuck with the basic design for two decades—far too long. Ford could embrace change in process, but not in product.

In 1921 some of his dealers asked Ford if he would vary the color of the Model T. His now famous answer was "You can have them in any color you want, boys, as long as they're black." Ford's competitors, however, who had begun to adopt his production techniques, were more responsive to customers. They gained ground steadily as sales of the Model T dropped. In May 1927 Ford announced that he would build a new car, the Model A. The Model T was dead, as was Ford's domination of the automobile market.

Ford's competitors realized the obvious benefits of interchangeable parts and a moving assembly line, but they did not grasp the power of the hidden paradigm—his systematic approach to time, as well as his systematic efforts to reduce resource use. For decades Ford's lessons were lost on American industry—particularly his own beloved auto industry—and they would have to be rediscovered, and refined, by the Japanese.

SHIGEO SHINGO AND TAIICHI OHNO

The Japanese have never been ungenerous about giving credit to Henry Ford. In 1982 Toyota President Eiji Toyoda told visiting Philip Caldwell, then head of Ford: "There is no secret to how we learned to do what we do, Mr. Caldwell. We learned it at the Rouge."[14]

Why were so many American businesses slow to embrace a systematic approach to production? Why were Japanese businesses faster? I believe there are two main reasons. First, Japan was open to change when W. Edwards Deming, and Joseph Juran and Armand Feigenbaum, began preaching a revolutionary message about quality in the 1950s. Japan had been devastated, both physically and psychologically, by the American triumph in World War II. Its products were notoriously shoddy. "Made in Japan" was a joke.

In 1950 Deming first presented to Japanese managers a "systematic approach to solving quality problems." He introduced a rigorous approach to consumer research and urged top managers to become involved in quality improvement. Eventually Deming became a national hero in Japan. Juran arrived in 1954, focusing on management, planning, and organizational approaches to improve quality. At the same time the work of Feigenbaum, head of quality at General Electric, became widely disseminated in Japan. Feigenbaum argued for "a systemic or total approach to quality (which) required the involvement of all functions in the quality process, and not simply manufacturing."[15]

American companies were exposed to these same ideas. But America had triumphed in World War II. What need was there for us to change, to improve? Weren't we as good as anyone could get? In the two decades that followed, American companies had little serious competition from foreign firms only slowly recovering from the war. Again, where was the motivation to change?

A dramatic paradigm shift requires dramatic evidence of failure. That would be some time coming for most American companies.

The second reason Japanese companies were able to embrace systems thinking in production is more fundamental. It is best told by Shigeo Shingo and Taiichi Ohno, who developed the Toyota production system in the 1950s and 1960s. Their views are worth noting not merely because they were the first to perfect fast-cycle manufacturing in Japan but also because they made such a good product. The Toyotas they built may well have been the best Japanese cars of their time, the best of the best. In a 1993 broadcast Tom and Ray Magliozzi, the auto mechanics of National Public Radio's *Car Talk,* called the 1976 Toyota Corolla "probably the most reliable, durable car ever made, including the Model-T Ford."[16]

According to Shingo, although "many people have thought about improvement . . . it was [Frederick] Taylor and the Gilbreths—F. B. Gilbreth and his wife, Lillian—who, in the 1890s, developed a clearly defined notion of improvement and established techniques to achieve it." Shingo notes that both Taylor and the Gilbreths focused on "time." He summarizes their work as follows:

- *Taylor:* Define the status quo analytically and temporally, and improve it through scientific reasoning—these activities are known as *time-study techniques.*

- *The Gilbreths:* Carry out motion analysis by breaking up the status quo into elemental units of motion called therbligs. Identify the purpose of each therblig, and find the one best way (in which work is broken down, purposes are tracked down, and better methods are devised) using techniques that accord with those purposes.[17]

Shingo and Ohno were heavily influenced by the American focus on time. Shingo, however, is very critical of how American companies applied the teachings of their own management experts. He wrote in the mid-1980s:

> Although Taylor argued for the establishment of standard times based on time analysis, many U.S. producers, in response to pressure to establish standards, set times without studying time. They simply accept the status quo (the times required for current operations) and then set mean standard times. . . . Conceptual approaches and activities basic to work improvement are often forgotten. . . . It seems that American management has lost sight of the original goal of doing away with waste.[18]

Means and ends had become disconnected in the United States.

A very similar point was made in 1988 by the American authors of *Dynamic Manufacturing*, who noted that Taylor "assumed that learning about production (through staff analysis) and the actual making of things (by the line organization) were separable activities: both could and should be done by specialists." The disconnection between learning by staff and making by line personnel, between thinking and doing, had serious results: It "increased the likelihood that learning only took place as a result of analysis that was divorced from hands-on experience, performed by people who were both physically and psychologically removed from the workplace."[19] It all seemed so much more efficient, and everyone was making so much money, that few noticed the fault lines.

Shingo offers an even deeper explanation for why the Japanese, but not the Americans, took a systems approach to production.[20] He begins by retelling the story of the British Industrial Revolution as related by Adam Smith in his 1778 book, *The Wealth of Nations*. Before the Industrial Revolution began in England in the mid-eighteenth century, the individual artisan made an entire product. The most skilled and diligent craftsperson could, for instance, make only twenty pins a day. Pin production increased 200-fold when labor was divided into eighteen separate tasks (such as cutting the material, sharpening the tip, attaching the head). The price of pins dropped considerably. At the same time unskilled laborers previously unable to participate in industry could now be paid for performing useful tasks. Demand exploded because far more people could now buy pins. This spread to other products and led to the rapid growth of British industry.

The "secret of the high productivity" gained through division of labor, Shingo writes, is that "simplified tasks made individual judgments unnecessary," and "the elimination of intermediate motions enabled work to be performed reflexively." Thinking was disconnected from doing:

> Until the Industrial Revolution, process (the flow from raw material to finished product) and operation (the flow of tasks performed by human workers on products) had been fused in the work of single individuals. One consequence of the division of labor was to separate processes from operations.[21]

The distinction between *process* and *operations* is critical to understanding the systems approach. *Process* refers to the "stages through which raw materials gradually move to become finished products": Worker A cutting a piece of raw material, Worker B sharpening it, and Worker C attaching the head. *Operations* refers to the discrete actions of an individual laborer working on different products. Worker B sharpening the first pin, then sharpening the second pin, and so on.[22]

To Shingo, process represents a frontal view of production, while operations fulfill a secondary function. But even though of lesser importance, an operation is performed in one spot and involves actually shaping the product. Operations are easier to see and control, and people "inevitably became captivated by directly observable human motion—that is, operational movements." Process phenomena, though far more important, end up escaping our attention. The result has been "the delusion that production is synonymous with operations."[23]

Although the division of labor clearly separated processes and operations, "this fact utterly escaped notice for some 170 years." To Shingo, "It wasn't until 1921 that Frank B. Gilbreth reported to the American Society of Mechanical Engineers (1) that production included process phenomena and (2) that processes were composed of processing, inspection, transport and delay, and storage." Gilbreth, however, committed a major error, according to Shingo:

[Gilbreth] claimed that processes were phenomena identified by analyzing production in large units and that operations corresponded to
small units of analysis. In other words, he saw processes and operations as phenomena lying on the same axes—phenomena that dif-
fered only in size of analytical unit. Despite his acknowledgment of
process phenomena, this error once again led process functions to be
buried in operations functions.

The effects of this error were, to Shingo, enormous. Many in the
West ended up thinking that overall production would improve
whenever operations were improved. The West was missing the forest for the trees, a flaw that can be seen in three examples from a
1990 collection of Shingo's writings:

• An automated warehouse is an *operations* improvement: It speeds
up and makes the operation of storing items more efficient. Eliminating all or part of the need for the warehouse by tuning production
better to the market is a *process* improvement.

• Conveyor belts, cranes, and forklift trucks are *operations* improvements: They speed and aid the act of transporting goods. Elimination
of the need for transport in the first place is a *process* improvement.

• Finding faster and easier ways to remove glue, paint, oil, burrs, and
other undesirables from products are *operations* improvement; finding ways not to put them there in the first place is a *process* improvement.[24]

These examples show that process improvements do not merely
eliminate unnecessary operations, critical though that is; they invariably reduce environmental impact. Reducing warehouse space
reduces the need for energy to heat, cool, and light it. Reducing
transportation reduces fuel use and exhaust fumes. Eliminating
"undesirables" means no glue, paint, oil, or scrap; it also avoids the
resulting cleanup and disposal.

Streamlining an unnecessary operation is little different from
downsizing a flawed company: The flaws remain, and no systemic
problems are solved. Eliminating those unnecessary operations is
the only way to get rid of the flaws. Harvard Business School professors Robert Hayes and Kim Clark noted in 1985 that in a

highly complex factory environment two approaches are possible: "One can either attempt to develop a highly sophisticated . . . information and control system to manage all this complexity, or one can set about reducing the complexity." The "American mentality," unfortunately, has "kept us from exploring the impact of changing the basic structure of problems."[25] The American mentality has also made it difficult for us to understand what the Japanese were doing right. The Appendix gives some examples of how U.S. companies tried to imitate Japanese success but failed, primarily because they focused on operations rather than processes.

Process is more than an incoherent collection of operations; the whole is greater than the sum of its parts. These points may seem obvious, and Shingo himself considered them self-evident. Since 1955 he "emphasized a theory of production management rooted in the assertion that production is a network of processes and functions. Indeed, it is probably fair to say that this is the basis of Taiichi Ohno's Toyota production system, a system that is arguably the first in the world to place major emphasis on process functions."[26]

TAIICHI OHNO AND THE FIVE *WHYS*

The man credited with inventing—or reinventing—the just-in-time production system is Taiichi Ohno. In the Toyota production system he developed after World War II with Shigeo Shingo, he could manufacture a variety of cars in small batches with the same processes. Their system made use of close supplier relations, total quality control, simplified production flows, and a scheduling mechanism that allowed employees to make decisions on the factory floor. Just-in-time allows Toyota assembly lines to stock only about two hours' worth of parts inventory, compared with two weeks' worth of expensive inventory for a typical General Motors plant.

Ohno studied Henry Ford's work and quoted his writings at length.[27] Like Ford, Ohno was obsessed with eliminating waste. Just as Ford devoted a whole chapter to learning from waste, Ohno wrote, "To implement the Toyota production system in your own business, there must be a total understanding of waste. Unless all sources of waste are detected and crushed, success will always be just a dream." Ohno reduced his formula for success to one line, which essentially defines waste as the difference between current capacity and the actually output of work:

Present Capacity = Work + Waste

Ohno was not talking about wasted resources. For him waste meant practices such as overproduction, waiting, making defective products, and excess inventory ("the greatest waste of all")—all of which were (and often still are) seen as standard business procedure, not as waste. Rooting out such waste is by no means easy. As Ohno wrote, "Underneath the 'cause' of a problem, the *real cause* is hidden. In every case, we must dig up the real cause by asking *why, why, why, why, why.*"

REPEATING *WHY* FIVE TIMES

1. *Why* did the machine stop?
There was an overload, and the fuse blew.

2. *Why* was there an overload?
The bearing was not sufficiently lubricated.

3. *Why* was it not lubricated sufficiently?
The lubrication pump was not pumping sufficiently.

4. *Why* was it not pumping sufficiently?
The shaft of the pump was worn and rattling.

5. *Why* was the shaft worn out?
There was no strainer attached, and metal scrap got in.

Failing to ask *why* five times might lead one simply to replace the fuse or the pump shaft, in which case the problem would recur within a few months. Ohno writes, "The Toyota production system has been built on the practice and evolution of this scientific approach. By asking *why* five times and answering it each time, we can get to the real cause of the problem."

Once again this essential idea came from America. It was Frank

Gilbreth, at the turn of the century, who argued for tracking down the goals of work by repeatedly asking *Why?*

Asking the tough questions and ferreting out the answers require participation by everyone, especially the production line worker. The manager is unlikely to know the right question to ask, and is even less likely to know the answer. And even the best company can improve. Perhaps that is why in 1982 Toyota's employee suggestion system received 1,905,682 suggestions (more than 32 per worker), of which 95 percent were implemented.[28] The commitment to improve the way things are done must be constant and must include everyone.

WHY CLEAN PRODUCTION AND GREEN DESIGN IMPROVE PRODUCTIVITY

The Japanese achieved their dramatic productivity gains by eliminating waste through improving the production process. Focusing on individual operations, what is called industrial engineering in this country, rarely achieves significant or enduring productivity gains. A narrow focus on operations makes it difficult to track down the root causes of systemic problems, such as long cycle time, large inventories, high waste, poor communications, flawed strategy, poor design, low quality. Yet systemic problems are the obstacles to productivity growth in most companies.

Pollution prevention is a key that unlocks the solutions to many systemic problems. Clean production, which in its highest form is pollution prevention, forces a manufacturer to look at the whole production process rather than only at operations. It forces a company to use many of the same techniques for minimizing wasted resources that Shingo and Ohno used to minimize wasted time, including cross-functional teams. Similarly, green office design (which is also pollution prevention) forces a service company to look at what it does systematically. Green office design also requires a team-based design. It leads that team to focus on the end users, office workers, and to improve control over their work environment. These are all essential elements of fast-cycle management.

Many businesses have been frustrated in their attempts to emulate Japan's success with lean production and Total Quality Man-

Clean production and green design lead companies to eliminate waste systematically by improving processes. That approach invariably increases productivity.

agement, mainly because they do not know how to take a systems approach to improving processes. This weakness results from the traditional U.S. focus on operations—a tradition that is one reason so few discuss systems thinking. Nevertheless, systems thinking is essential for achieving significant and enduring gains in profits and productivity and lies at the heart of lean and clean management.

2

SYSTEMS:
THE CYCLE OF LIFE

Symbolic analysts—who solve, identify, and broker new problems—
are, by and large, succeeding in the world economy. . . . The education of the symbolic analyst emphasizes systems thinking.
— ROBERT REICH, *The Work of Nations*[1]

Few businesses view their facilities as a source of moneymaking
investments. Yet Ron Perkins, facilities manager for Compaq
Computer in the 1980s, saved his company about $1 million a year
by reducing energy use (and hence pollution). To do so, however, he
had to overcome a major obstacle blocking change: In the buildings
and facilities division of Compaq, as in most companies, projects
had traditionally required a payback period of two years or less. If
an investment took more than two years to pay for itself, forget it.[2]

In an innovative move, Perkins went to Compaq's chief financial
officer, John Gribi, one of the few people he knew at the company
who had a long-term view. Gribi's philosophy was simple: *Apply
capital dollars to reduce future expense.* In other words, "Spend
money to save money." With that proactive mission in mind, Perkins went to the treasurer to get capital for his money-saving projects. The treasurer had come from a top accounting firm and, according to Perkins, "didn't even know what payback was." He told
Perkins, "I don't do that. I do return on investment [ROI]. I compare

the ROI with the company's cost of money [basically, the interest rate the company pays to borrow money], which ranges from 7 to 11 percent. Give us corporate cost of money plus 3 percent. If we don't have the money, we'll borrow."

In other words, Compaq would make almost any investment in itself with an ROI of 14 percent or more.

THINK LIFE CYCLE

By connecting the finance people with the facilities people for the first time at Compaq, Perkins had shattered the traditional mind-set of the two-year payback, which demands a 50 percent return on investment. The new goal was a five- to seven-year payback. Perkins notes, "That's how we were able to do life-cycle costing, to figure out the effect of any change on the ability of the company to make profit." Perkins now knows that in judging any proposed capital improvement, he can look five to seven years out and make money for his company.

Perkins had revealed the connection between the company's capital spending on its buildings and its future operating costs. A life-cycle analysis of any purchase considers its total cost over many years, including the money needed for energy use, operations, and maintenance. Compaq could spend a little more now to save a lot more later. Initial costs might be a bit higher, but life-cycle costs would be far lower. Most of Perkins's improvements, such as daylighting Compaq's buildings, were also designed to improve worker productivity (see Chapter 4), or, as Perkins puts it, to "systematically remove the barriers to productivity." Productivity growth at Compaq was so rapid—55 percent in 1985 alone—that it is not possible to trace it to any one cause.

Ron Perkins succeeded at Compaq by using systems thinking, which requires making new connections in space and time. In his search for money to reduce energy use, pollution, and long-term operating costs, he helped break down not only the barrier between facilities and finance but also the barrier between the company's present purchases and its future costs. Compaq's people became more physically interconnected and more forward thinking—two hallmarks of systems thinking.

Life-cycle analysis lies at the heart of a systems approach to becoming lean and clean. It leads to energy efficiency, as Perkins demonstrated, and to high-productivity workplaces, as well as being crucial to designing green products and clean production processes. Overuse of resources at the beginning of the production process leads to excess pollution at the end. A company can achieve huge savings at both ends by reducing resource use and pollution when a product is manufactured, when it is used, and when it is ready to be discarded. Again, a life-cycle approach is needed to minimize overall costs, not merely initial costs. This in turn requires a systematic approach to production: The people who research, design, manufacture, and market must work together at the same time in cross-functional teams.

In sum, becoming lean and clean through systems thinking requires making connections in time (through life-cycle analysis) and in space (through teamwork). Systems thinking is the only way to achieve process improvements, which is the only way to raise productivity significantly, which is the only enduring way to achieve success in business. For that reason, and because, as Robert Reich has noted, those who understand systems thinking are succeeding in the global marketplace, it is worth examining the principles of systems thinking and how they apply to lean and clean management.

WHY SYSTEMS THINKING WORKS

What is a system? Why is systems thinking so powerful? How can it help us see and do the obvious? Why is it an essential part of lean and clean management? One of the goals of this book is to answer these questions.

The simple definition offered by systems dynamics expert Donella Meadows is a good start: "A system is any set of interconnected elements."[3] The emphasis is on connections and relationships.

Interconnections are the essence of the ecosystem. One creature's waste is another's food. Nature squeezes every last bit out of its resources. Our natural environment is a system, and dealing with it properly requires a systems approach. The word *environment,* after all, comes from Old French, *viron,* meaning "circle." Since the word *cycle* also derives from *circle,* the environment could be called the cycle of life. Ultimately, businesses will learn the lessons of nature, to squeeze every last bit out of their resources and create an *industrial ecosystem,* where one company's waste is another's resource.

Systems thinking is not easy. As Peter Senge, director of the Systems Thinking and Organizational Learning Program at M.I.T.'s Sloan School of Management, wrote in his book *The Fifth Discipline:* "Reality is made up of circles, but we see straight lines. Herein lie the beginnings of our limitations as systems thinkers."[4]

Systems thinking is counterintuitive for many of us because it is "backwards" thinking. You begin at the end. Rather than provide the same old goods and services, you start with the desired end state—such as giving the customer what he or she wants—and work backwards. Or, as Perkins did at Compaq, start with a focus on lowering future costs and work backwards to find the best investments today to do so.

If a systems approach is so powerful, why do so few preach and practice it? A key reason is that what passes for systems analysis at most companies is not systematic, so it cannot and does not work. Senge explains the source of the problem. Traditional systems analysis fails for the same reason that many tools of forecasting, business analysis, and strategic planning fail to achieve dramatic results for businesses: "They are all designed to handle the sort of complexity in which there are many variables: detail complexity." But Senge identifies a second type of complexity, "dynamic complexity, situations where cause and effect are subtle, and where the effects over time of interventions are not obvious."[5]

The crucial difference between detail complexity and dynamic complexity is their approaches to time. All processes involve time, but systems thinking makes explicit an emphasis on the long-term view. It seeks to understand the unexpected or unintended consequences of today's actions. Following a complex set of instructions to assemble a machine involves detail complexity. Designing a machine that is easy to assemble in order to reduce overall product cycle time, and easy to recycle when its life is over in order to reduce pollution, involves dynamic complexity. Taking inventory in a discount retail store, for example, involves detail complexity. Designing a retail system that minimizes inventory and rapidly detects and responds to changing customer purchase habits involves dynamic complexity.

Many firms can handle the detail complexity of analyzing the initial costs of materials, labor, and equipment to minimize short-term manufacturing costs and maximize short-term profits. Far fewer can handle the dynamic complexity behind determining the benefits of minimizing the life-cycle cost of product and process design and avoiding waste and pollution. Fewer still can determine the long-term payoff of training workers to minimize life-cycle costs. The designers, workers, and resource and environmental costs are all connected in dynamically complex ways.

The methods developed in the years after World War II may have been adequate when our global competition was still recovering from the devastation of the war, when financial markets and resource prices were more stable, and when environmental concerns were easily ignored. But in today's world of frenetic global competition and increased environmental concern, most American companies are frustrated to find that the methods they have relied on for decades can solve few of their problems. Methods designed for a static business environment are relatively useless in a dynamic global economy. Senge concludes, "The real leverage in most management situations lies in understanding dynamic complexity, not detail complexity."[6] Unfortunately, U.S. management tends to be static rather than dynamic and focused on the short term rather than the long term, so even promising new approaches, such as Total Quality Management (TQM), are bound to fail.

The most important focus for a company seeking a systems approach is *time* rather than *quality*. This counterintuitive notion is a fundamental limitation of most approaches to TQM. Time might seem to be the enemy of quality. We need to examine time anew. Time, particularly product cycle time, is inherently a dynamic concept. Quality isn't. Improving quality remains essential. It reduces cycle time, as the Japanese have shown. Making something right the first time is invariably faster and more efficient than trying to fix it later. But TQM has been too narrowly named.

Of all the possible ways to improve productivity, only just-in-time is statistically, demonstrably effective (see Chapter 7). Quality is an important goal, but by making it the central focus, TQM tends to achieve only that one goal—improved quality—and misses the additional, myriad gains made possible by focusing instead on time: reduced inventories and overhead, fewer bottlenecks, improved process, increased productivity, and, most important, faster response to the market and to all change. A narrow focus on quality dooms TQM.

Japanese companies achieved some of their most remarkable results in lean production by first adopting the focus on quality advocated by U.S. experts and then improving it to focus on product cycle time. They gave us fast-cycle management. A traditional, compartmentalized company responds too slowly to changing market conditions; the design crew learn too late that their product is too hard to make, and that the customers won't like it. As before, a systematic approach to time in an organization requires a systematic approach to space—getting everybody together and formally interconnected. A company must become a "set of interconnected elements," as Donella Meadows put it.

A focus on time, on the future impact of present actions, also leads to life-cycle analysis, which examines total costs over time rather than merely initial costs. The Golden Rule of lean and clean is "Think life cycle." Life-cycle analysis leads to energy efficiency and clean production. Thus, focusing on time leads inexorably to lean *and* clean management.

THE FRUITS OF SYSTEMATIC PROCESS REDESIGN

The companies that have succeeded in raising productivity with Total Quality Management and lean production have done so because they apply systems thinking to the production process to minimize wasted time. *Pollution prevention raises productivity because it forces companies to apply systems thinking to the production process to minimize wasted resources.*

Consider the Regal Fruit Co-op in Tonasket, Washington, which stores 1.25 million bushels of fruit for farmers in the north-central part of the state. Apples and pears have traditionally been stored at

31° to 32° Fahrenheit in an atmosphere of almost pure nitrogen, with the facility's fans running twenty-four hours a day. In 1989 Bonneville Power Administration approached Regal Fruit with a proposal to save energy and prevent pollution. Researchers at Washington State University had found that an astonishing 60 percent of the heat in an apple-storage facility came from the fans. Waste was designed into the process—a classic systems problem. But with a computer-controlled monitoring system, the fans needed to be on only about six hours a day, in a repeated cycle of two hours on and six hours off. In other words, the motors that run the fans could be turned off 75 percent of the time.[7]

The new equipment cost $104,000, more than half of which was covered by Bonneville Power. Yearly electricity savings were estimated at $13,000. Measured energy savings exceeded estimates by $2000 to $5000 a year. Regal Fruit's investment would pay for itself in three to four years, and, for one warehouse, electricity savings exceeded 350,000 kilowatt-hours a year.

More important, however, productivity went up, as is often the case when a company reexamines and redesigns a process that has been used for many years. The researchers had found that by reducing the oxygen content they could not only increase the temperature a few degrees (saving more energy) but also improve the quality of the fruit.

Fruit is typically stored five to nine months before being taken to market. "We used to factor in a shrinkage rate of up to 20 percent," says Ron Gonsalves, refrigeration supervisor for Regal Fruit. "With our Golden Delicious, we raised the temperature to 33 degrees, from 31, eliminating the defrost cycles, which contributed to dehydration. Now our shrinkage is only 1 percent." Whereas Regal once might have sold 100 partially dehydrated apples in every bushel, now they can sell 85 plump, juicy apples.

The savings from reduced shrinkage "far exceeds anything we make back from energy savings," says Gonsalves, "probably tenfold." In other words, while the energy savings would have paid for the new system in three to four years, the productivity rise paid for the system in less than one year. The goal of resource efficiency and pollution prevention led to systematic process redesign that increased productivity. That is environmental reengineering.

Although utilities, universities, and government are doing more and more to help industry, most companies will probably not have the benefit of outside assistance in redesigning a basic process or in systematically preventing pollution through resource efficiency. For that reason, companies must learn the elements of systems thinking so that they can become lean and clean on their own.

Embracing a dynamic, systems approach to management requires adopting practices that embed a time-based approach into everything a company does, practices that lead inevitably to redesign of both products and processes. Three basic management principles lie at the heart of systems thinking:

THE THREE KEYS TO SYSTEMS THINKING

1. Be proactive

2. Focus on the end results

3. Improve constantly

The most effective people practice systems thinking (often without knowing it). No other approach can achieve enduring success in an increasingly fast-paced and interconnected world.[8]

1. BE PROACTIVE

Being proactive, as opposed to reactive, means initiating action, not waiting for events to force you to act. Being proactive is dynamic, not static. It is forward looking, driving you to think of the future consequences of today's actions. The best way to be proactive is to anticipate problems and prevent them.

Focus on Prevention

Prevention is as essential to clean production as it is to lean production and TQM. Lean and clean design of product and process prevents pollution and avoids the tremendous costs of dealing with pollution after the fact—just as designing quality into product and process prevents defects and the tremendous costs associated with fixing flaws after the fact.

Train Workers

The best way to prevent job loss and pollution is to train workers in lean and clean techniques. As explained later in this chapter, by training its workers in prevention, Dow Chemical reduced its energy, resource, and pollution costs by millions of dollars a year. Don't let their skills become outdated. Train workers to think proactively and focus on prevention, to think constantly about improving process and product, and to think above all about customers.

Proactive measures have a high return on investment. Investments in pollution prevention and training have rates of return for some companies of 50 percent to 300 percent. Being reactive produces very low, if not negative, returns on investment, because the damage has already been done. Once created, pollution of any kind—air pollution, dirty water, toxic waste—is very expensive to deal with. Costs to a company may include handling (and its possible health effect on workers), disposal, treatment, legal costs, and bad publicity. Preventing fires may not be as glamorous as putting them out, but it is far more cost-effective.

2. FOCUS ON THE END RESULTS

Measure Results

Unless a company measures the results of its actions, it will never know whether changes have a positive or negative effect. That is why successful companies, whether in the manufacturing or the service sector, continually track service delivery time, measure product quality, survey customer satisfaction, and, increasingly, monitor their pollution. *The Wall Street Journal* noted in a June 1991 article on pollution prevention, "In a major shift, chemical companies are viewing waste not as an unavoidable result of the manufacturing process, but as a measure of its efficiency."[9]

Talk to Customers

The end users of a product are customers, and they alone determine what quality is. Talking to customers is crucial to lean and clean management. The chairman and CEO of Procter & Gamble, Edwin L. Artzt, told his shareholders in 1990, "We hear from environmentally aware consumers every day on our toll-free phone lines and through the mail—telling us to keep reducing the amount of packaging we use and to use recycled or recyclable materials whenever we can."[10]

A company ignores its customers' desires at its peril. Customers' desires are never obstacles; they are always opportunities for meeting a new market niche and for developing a new environmental partnership. That's Xerox's philosophy, as proclaimed in a recent company case history:

> Recognizing customer concerns about disposable products, Xerox has developed procedures for retrieving and recycling customer-replaceable copy cartridges. The company has now established design standards for its future products, including copy cartridges and toner containers, that provide an integrated approach to extended life, reduced cost, recycling, and remanufacturing compatibility. Customers participate in partnership with Xerox, returning copy cartridges.[11]

Talk to Employees

Employees are the end users of a company's office equipment and buildings. Find out what makes them happy and give it to them. For example, people love daylight; it makes them more productive. Daylighting has extra benefits: It saves energy, cuts costs, and prevents pollution.

A focus on the end users—the customers and line employees—inevitably leads to a *least-cost approach*. For example, energy guru Amory Lovins noted in 1977 that people don't want electricity or oil, they want "comfortable rooms, light, vehicular motion . . . and other real things."[12] By focusing on the end user, he found that it was possible to have better lighting and more comfort with far *less* energy use—50 percent to 90 percent less. Comfort and adequate illumination are in the eye of the beholder. The point may seem obvious, but many offices have windows that cannot be opened by the occupant. And many offices are still lit for writing and typing on horizontal surfaces and thus are mislit and overlit for using computers. Focusing on the end result in office design—designing a good work environment—is the least-cost approach to raising worker productivity.

Giving people what you think they want, or what you have always given them, costs you more than you think. It wastes the difference between what you supply them and what they actually want. Finding out first what they want, then giving them only that, avoids the waste. They are happier and costs are surprisingly lower.

Reducing waste is another central goal of a systems approach to lean and clean management. Energy efficiency reduces wasted resources by better delivering the energy services demanded by employees and customers; when practiced by electric utilities, energy efficiency is often called demand-side management. Lean production eliminates wasted time by focusing relentlessly on the end user. Quality is never achieved by simply trying to reduce defects per thousand but rather by working with customers to give them what they want, which inspires everyone in the company to make it right the first time. Achieving just-in-time inventory control (as opposed to the traditional just-in-case approach) requires designing a production process in which individual components are built or provided only as they are demanded by the next phase in production (called demand-pull manufacturing). Ultimately, people want products, not pollution. Clean production aims to deliver the former without the latter.

3. IMPROVE CONSTANTLY

Make change slowly, but steadily: To be systematic about time means to be dynamic rather than static, to anticipate the future, to accept that change is never ending. The only companies that will succeed are those committed to constant change, to continuous incremental improvements.

The Japanese term for this principle is *kaizen* (ky-zen): *ongoing* improvement involving *everyone*—top management, managers, and workers. According to management consultant Masaaki Imai, "Kaizen strategy is the single most important concept in Japanese management—the key to Japanese competitive success."[13]

Lean production is the powerful embodiment of kaizen. According to the M.I.T. Commission on Industrial Productivity, "The cumulative effect of successive incremental improvements and modification to established products and processes can be very large and may outpace efforts to achieve technological breakthroughs." American companies have often lagged behind foreign competitors in exploiting the benefits of "continuous improvement in the quality and reliability of their products and processes."[14]

Occasionally, change will have to be fast. If a company is completely on the wrong track, it may have to be "reengineered," to use the current management consultant lingo.[15] But the shift to lean and clean can be stalled if change is too rapid, and for most corporations the ultimate goal should be slow but steady change.

CONTINUOUS ENVIRONMENTAL IMPROVEMENT: DOW CHEMICAL

Is continuous improvement a concept that really applies to energy efficiency and pollution prevention? Absolutely. Consider Dow Chemical's Louisiana Division, which has 2400 employees in over twenty plants making chemicals such as propylene.[16] The division began a yearly contest in 1982 to find energy-saving projects that paid for themselves in under one year (an ROI of over 100 percent) and that required a capital investment under $200,000.

The first year had twenty-seven winners requiring a total capital investment of $1.7 million with an average return on investment of 173 percent. Dow's energy manager, Kenneth Nelson, reported that after the first twenty-seven projects "many people felt there couldn't be others with such high returns." The skeptics were wrong. The 1983 contest had thirty-two winners requiring a total capital investment of $2.2 million and a 340 percent ROI—a savings for the company of $7.5 million in the first year, and every year after that!

The yearly contest was so successful that the rules were changed to eliminate the $200,000 limit and to include savings from increased production yield or capacity. Gains from all kinds of waste reduction (not just energy) would count. Some of the winning projects were simple energy-saving ideas such as more efficient pumps, or use of heat exchangers to salvage waste heat. Others involved more sophisticated process redesign of the kind discussed in Chapter 8. To foster savings, Dow published a "Waste Elimination Idea Book" and set up a two-and-a-half-day Continuous Process Improvement Workshop.

The results of most projects are measured through detailed audits to determine "actual project costs, the pounds or BTUs saved, and anything new that was learned." For 575 projects with an average return of 202 percent (based on the submitted proposal), the audited return was 204 percent—the projects actually performed a little better than expected. All projects are reviewed, and employees with winning projects receive engraved plaques in a formal awards ceremony. Although there is no direct cash award, many of the winners are rewarded "by their own supervisors through normal channels" (i.e., they get raises based on total job performance).

Even as fuel prices declined, the savings kept growing. The best year was 1989, when sixty-four projects costing $7.5 million saved the company $37.0 million in the first year, and every year after, for a return on investment of 470 percent—a payback of under three months. Even after ten years, and nearly 700 projects, the 1992 contest had 109 winners with an average return of 305 percent, and the 1993 contest had 140 projects with an average return of 298 percent.

This is pollution prevention kaizen.

THE GOLDEN TOOL OF SYSTEMS THINKING—FEEDBACK

Continuous improvement and focus on the end result both require constant feedback: rigorously measuring results and rapidly reporting them to the right people. Since most companies aren't systematic about feedback, they miss most of the potential gain. Many companies have their waste handled by their environmental department, so the cost of pollution does not feed back to those who create it: the research and development and manufacturing departments. Without feedback those departments have little financial incentive to incorporate pollution prevention and process efficiency into their planning. Hewlett Packard solved that problem by distributing waste management costs directly to manufacturing cost centers.

Similarly, being proactive—focusing on prevention—is rare because there are so few institutional rewards for it. You see a leak in an underground crack. You could fix it now, but that would cost

$10,000. You're already overwhelmed with work, you're not in the fixing business anyway, no one would reward the effort, they may even label you a troublemaker, so you move on. In April 1992, in Chicago, that crack burst, forcing the evacuation of tens of thousands of people and costing more than $1 billion.

Plugging the leak costs real money now. The savings—though enormous—never appear if the disaster never occurs, so no one can be rewarded. Without positive, reinforcing feedback, how can the desired behavior be reinforced? The solution is to embed prevention in corporate culture. Companies need to train workers to prevent defects, prevent pollution, and prevent machine breakdown. More important, companies need to reward the group with the fewest defects, the most recycling, the least waste, the least downtime, the fewest serious cracks. That will in turn require measuring those results.

Feedback is the quick and constant return of information to you about the consequences of what you are doing. The best way to reinforce desired behavior, or discourage undesired behavior, is to increase feedback.

Why do individuals, groups, and nations remain stuck in counterproductive behavior? Why do they fail to embrace new thinking? Because the consequences of the flawed behavior are distant in both space and time. The greater the distance, the more difficult it is to prove cause and effect and the more we must rely on some abstract theory to explain how our actions affect the world.

We rarely overfill a glass of water; we can see the water level directly, and our control of the faucet is nearly instantaneous. We often overheat the shower water because of the time lag between the turning of the hot water knob and the increasing of the temperature of the water coming out of the showerhead. We overpollute the planet because the most damaging effects of the overuse or misuse of resources are seldom either nearby or immediate. Using an inefficient lightbulb in the office causes the release of unnecessary air pollutants at a fossil-fuel power plant miles away, which in turn contributes to acid rain hundreds of miles away, which might in decades help destroy a forest or a lake. Were it not for the tremendous savings possible today from improved lightbulbs and other energy-efficient measures, who would ever see the need for change?

The Power of Feedback

There was once a new housing development in the Netherlands where, by accident, the electric meters were installed in the basements of some houses and in the front halls of other, otherwise identical, houses. Electricity use in the houses where the meters were easily visible was 30% less than in the houses where the meters were out of sight. The only difference was feedback.

—Donella Meadows, 1991[17]

When companies had no feedback about their use of chlorofluorocarbons (CFCs), they had no reason to change. Then the world learned that CFCs are destroying the ozone layer, which protects us from harmful ultraviolet radiation. As public opinion turned firmly against the chemical, companies began to shift their mind-set. Using the techniques of clean production, many found that CFCs were much easier to live without than anyone realized. To replace CFCs with new cleaning solutions, AT&T adopted a systems approach in the late 1980s and redesigned its cleaning processes. The result: The company cut the cost for cleaning circuit boards to fifteen cents a square foot, down from twenty-five cents, for a total yearly savings of $3 million.[18]

Much of the pollution we take for granted comes from inefficient processes developed decades ago, before we understood their true costs. The processes—and the corporate culture that justified them—came from a time when our abundance of resources and land made it easy to ignore the dangers of resource overuse and pollution. But times have changed.

With better feedback we can see the results of our actions instantaneously, and more rapidly abandon flawed thinking and determine a superior course of action. Imagine what would happen if cigarettes killed smokers in two years rather than forty. Medical researchers would have established the link faster, and people would find it far easier to quit smoking. And how much lower would America's monumental trade deficit with Japan in cars and car parts be if every Detroit manager in 1979 had had to spend one week working with garage mechanics on U.S. cars and Japanese cars?[19]

In any systematic approach to management, and especially lean and clean management, a basic tool is improving feedback. How can feedback be encouraged and increased in an organization?

Regularly talking to one's customers is the most important feedback device, since product quality is defined by the customer. The other elements of Total Quality Management are also powerful: fewer layers of management, an employee suggestion system, measurement of results, and cross-functional teams. These policies in turn speed up the product cycle. The faster the product cycle, the faster the learning process, and the more opportunities there are to learn from the feedback of experience.[20] Employees and managers can more quickly see what works and what does not. Thus, improved feedback accelerates the process, which in turn improves feedback. This itself is a classic example of positive, reinforcing feedback. It helps explain why, once a company gets on the right path, it can achieve remarkable results (whereas companies that fail to make the necessary systems changes fall further and further behind).

Positive feedback is crucial for reinforcing, accelerating, and institutionalizing change. When people cannot quickly see the benefits of their actions, their frustration and resistance will be more difficult to overcome. Dramatic reorientation requires rapid, positive feedback, for when people see the benefits of their actions in a short time, they are far more likely to commit to change. Unfortunately, the effects of improvements in quality and cycle time, such as increased market share, may take many months to become clear. Some of the most important benefits, such as increased learning, may take even longer to demonstrate. Some, such as the accident that doesn't happen or the customer that isn't lost, never become fully clear.

Fortunately, the benefits of good environmental practices, especially those focusing on energy use, are quickly evident. After a well-designed lighting retrofit, employees will see in a few short months a variety of improvements: Lighting electricity use is reduced by well over 50 percent, while saving money, improving lighting quality (and hence increasing productivity), and on top of all that lowering air pollution (plus winning an EPA seal of approval). When such remarkable achievements are accomplished so rapidly, when feedback is reinforced so dramatically, longer term goals, such as a 90 percent reduction in defects and 75 percent reduction in cycle time, will seem within grasp.

An essential goal of systems thinking is to instill permanently into a company's managers and workers a focus on time, on the future consequences of today's actions. We tend to ignore that future impact because there is no feedback *today* of that impact. Such feedback comes only long after we have acted, which underscores the importance of life-cycle analysis to systems thinking. Life-cycle analysis requires examination of all costs associated with any decision, not merely the obvious initial costs. It forces a company to consider the future impact of today's actions *today*. Life-cycle analysis institutionalizes the feedback. This is particularly important because *decisions made during the early phases of a project commit the vast majority of life-cycle funds*—long before there is any feedback.

The graph on this page shows how much of the life-cycle cost of a project is committed during its various stages.

When a mere 1 percent of the costs are spent on a project, 70 percent of its life-cycle costs are committed. By the time 7 percent

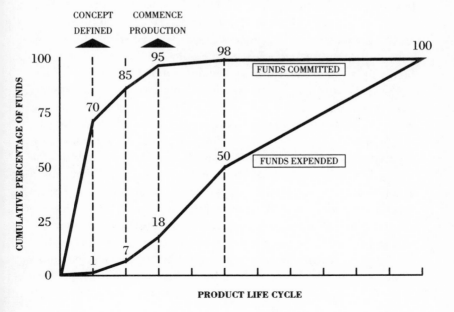

PRODUCT LIFE CYCLE

of the costs are spent, 85 percent of the life-cycle costs are committed.[21] The message of the graph: All the people who have useful knowledge must contribute to a project early on. Cross-functional design teams are crucial for lowering life-cycle costs, as we will see with both product design and building design.

Consider the Fisher Scientific plant in Fair Lawn, New Jersey, which employs 130 people to make over 1400 chemicals. A study of its pollution prevention efforts found that twenty-eight projects, costing a total of $79,000, cut waste an average of 43 percent while increasing production yield an average of 28 percent, saving the company $529,000 per year. Fisher had originally set up a waste minimization committee consisting of representatives from engineering, operations, and research to review all processes and proposed changes for yield improvement and source reduction opportunities. The limited membership of the committee, however, resulted in many false starts, since, for example, accounting and sales were not represented. An accountant could identify places where waste might be particularly costly, and someone from sales might know what outlets exist for selling wastes as by-products. The study noted, "Fisher officials said they learned that the only effective approach for their waste minimization committee is to have a full-spectrum multi-disciplinary team analyzing the problems."[22]

Since the decisions of a process design—or redesign—team have such a disproportionate effect on the outcome, team members had better be as knowledgeable and highly trained as possible. Life-cycle analysis of employees will inevitably lead to far greater expenditures on training. Treating employees as a depreciating asset, whose value declines with time, is a self-fulfilling prophecy, leaving a company with underskilled, undermotivated, and thus underproductive workers. Investing in employees' skill is like performing preventive maintenance on the workforce. No investment has a greater payback than training workers to prevent pollution, prevent machinery breakdown, deliver quality, and serve customers—in short, to perform life-cycle analyses and use systems thinking.

The single most important skill for high-wage jobs is the ability to understand systems, according to a 1991 Labor Department study. The study surveyed twenty-three high-wage jobs and twenty-three low-wage ones to determine which were the most important skills.

For each job the authors talked to both the workers and their supervisors. They determined how important it was that a worker be able to use *resources* (such as money), *interpersonal skills, information* (such as using computers), *technology,* and *systems* (understanding, monitoring, designing, and improving social, organizational, and technological systems). The study concluded that systems thinking is the skill that most distinguishes workers in high-wage jobs from those in low-wage jobs. And of the forty-six jobs surveyed, the Labor Department concluded that the one that required systems thinking above all other skills was "Blue Collar Worker Supervisor."[23]

I am writing here to managers *and* workers because both must master systems thinking and lean and clean management if their companies are to succeed. Successful fast-cycle businesses are giving more managerial responsibilities to line workers and relying on them to suggest continuous improvements to product and process. To perform as manager in part, workers will need a keen understanding of all aspects of improving profits and productivity. To delegate authority successfully, managers must have reason to be confident that workers share their leadership philosophy. The next chapter will focus on the essential role workers have in making a company lean and clean.

3

LABOR AND
LEAN AND CLEAN

Man can put out about 1/20 of a horsepower. He has to rest at least 9
hours a day. He also has to eat and drink. As a power source, we are
terrible. However, it is when man starts thinking of ideas that the
difference between man and machine emerges.

—SOICHIRO HONDA[1]

Workers have a special interest in lean and clean because it
offers the possibility of finding *nonlabor* cost savings: reducing energy use, waste, and pollution. Lean and clean increases productivity, cuts costs, and reduces pollution, but it does so *without
firing workers.* By becoming expert in lean and clean, workers can
offer a powerful alternative the next time management proposes cutting costs the traditional way—through layoffs and wage reductions.

From management's perspective, the change to lean and clean
should be seen as inevitable. Bringing employees into the process
from the outset is the best way to harness their commitment and
knowledge. Imagine a company that offered wage increases to its
workers if they could identify and implement significant nonlabor
cost savings. That company would be creating a powerful lean and
clean partnership between management and labor. Such a company
exists.

REPUBLIC ENGINEERED STEELS SAVES MONEY AND JOBS

Republic Engineered Steels uses employee brainpower to cut costs, reduce pollution, save jobs, and raise wages. Republic was bought by employees in 1989 from the failing LTV Corporation. They needed to cut costs $80 million a year to remain solvent.[2]

Of Republic's 5000 employees, roughly 4000 are unionized. After the employee buyout, "we realized we couldn't succeed with an adversarial relationship" between management and the union, said Harold Kelly, a company vice president. In 1991 they put in place Project 80, a program to reduce costs $80 million a year by encouraging employees to identify process changes that would eliminate waste. Employees identify an improvement, a management-union committee reviews it, workers develop an action plan and implement it.

In the first twenty months of Project 80, employees suggested about 1000 improvements, about half of which are being put in place. Some $63 million in savings have been identified, and $45 million have been achieved. The steel industry has had a tough time in the recession, and Republic laid off about 100 employees. The company estimates that, without the suggestion system, it would have had to let go more than 600 workers.

The single largest money-saving idea was for improved recycling of scrap steel, which cut more than $3.5 million off the budget. The environmental department has recycled more than 3000 steel drums. These are the fifty-five-gallon drums in which Republic's chemicals are delivered. The company had been paying thousands of dollars to remove the drums and dispose of them. Then the environmental department realized that the steel drums could be recycled. "So instead of paying someone to get rid of them, we're using them to make steel," says Kelly. Or, as Henry Ford said seventy years ago, "Instead of paying freight on waste, we keep the waste and earn money from it." Better still, eliminate scrap in the first place: At Republic's Gary, Indiana, plant, a quality team cut in-plant salvage 40 percent.[3]

More efficient use of water has been another huge money saver. One group of workers figured out a way to reduce water used during the heat-treating process. Another group proposed a water-efficiency program at the main plant in Canton, Ohio. The savings ranged from little suggestions (replacing two drinking fountains with watercoolers) to bigger things (such as fixing leaking pipes) and major efforts (recycling rinsing water). Water consumption has dropped 80 percent, from more than 9 million gallons a month in 1991 to under 2 million two years later. Savings are close to $50,000 a year.

Ray Milini, a seventeen-year veteran of the plant and one of the hourly workers who proposed the water-efficiency program, says of the suggestion system, "Before, they didn't want to know what I might tell them, so I wouldn't tell them. It makes you care about the product. You actually got a hand in it."

The cost cutting has not come at the expense of quality. In fact, many suggestions have improved quality. For example, the order department streamlined the process of filling orders, cutting the number of product returns caused by entry error from 116 to 38 a year. Republic invests heavily in quality. After two benchmarking trips, Republic built and installed a $12 million Quality Verification Line, which includes a magnetic flux leakage detector that spots surface defects as small as 0.008 inch deep by 0.625 inch long; an ultrasonic test machine that can detect internal defects as small as 0.020 inches; and magnetic resonance to verify that each bar has the specified steel content. Republic won one of the six Manufacturing Excellence awards from *Controls and Systems* magazine for 1991, and the Labor Department cited Republic in 1991 for its LIFT (Labor Investing for Tomorrow) America Award for "designing and managing innovative programs to improve the skills of American workers."

Republic does not give direct monetary awards for suggestions that it uses. But then Republic Engineered Steels is a special case: Its workers are its owners. So suggestions don't merely benefit them by saving their jobs; suggestions improve the return on their investment in their own company. Employee ownership is one of the best uses of feedback.

In 1993 Republic offered another incentive to workers for achieving nonlabor cost savings. The company would advance workers a raise against achieving several million dollars more in cost savings. Once those cost savings were achieved, Republic would advance workers another raise, and so on. This is a true partnership for lean and clean kaizen.

JAPAN: A GIANT SUGGESTION BOX

Any company seeking to harness workers' brainpower requires a systematic approach: an employee suggestion system. Though essential to a competitive company, this approach is rarely taken seriously and is often the butt of jokes and cartoons. It is not a butt of jokes at Republic or at most Japanese companies.

The Japan Human Relations Association traces the origin of the modern suggestion system to America (where else?) in 1898. In that year records show than an Eastman Kodak worker received a prize of two dollars "for suggesting that windows be washed to keep the workplaces brighter." A fitting omen—the first known modern industrial suggestion was how to improve lighting at low cost.[4]

Today a suggestion system is central to becoming lean and clean. The two main subjects for suggestions in Japan are (in order)

- Improvements in one's own work

- Savings in energy, material, and other resources

The Japan Human Relations Association offers the following motto for a suggestion system: "Limited resources, unlimited ideas."

Although the modern suggestion system was born in America, it, like many U.S. management practices, has been brought to a high art in Japan. Let's benchmark against our competition. The Japanese company Canon has a strong employee suggestion system because of its commitment to kaizen, constant improvement. In 1985 employees submitted almost 900,000 improvement suggestions, roughly the total number of suggestions submitted by all U.S. employees to their companies. Canon received 78 suggestions per employee. One employee submitted 2600. Canon paid out $2.2 million in prizes, but saved over $200 million as a direct result of the suggestions.[5] There are no jokes about suggestion boxes at Canon.

Don't be overly impressed by Canon—twelve other Japanese companies outperformed it. In first place Matsushita had 6.5 million suggestions that year. Let me repeat: 6.5 million employee suggestions in one firm in one year. On average, 76 percent of employee suggestions are used in Japanese companies. One Japanese company, Aisin Warner, which makes automatic transmissions, gets 127 suggestions per worker and uses 99 percent of them.[6]

How do Japanese companies generate such a remarkably high number of valuable ideas? They give rapid feedback—rewarding all suggestions used within a few days and explaining every one not used. In contrast, American companies commonly seek and accept only a few good ideas, wait weeks before giving a response, and often give no reward, even for suggestions that are implemented.

A typical Japanese employee suggestion is illustrated in the figure on page 55.[7] It saved 40 percent of cleaning time and $642 a month in energy costs. This lean and clean suggestion could have been thought of only by the person working on the machine day in and day out and asking, "How can I improve this?"

Shigeo Shingo writes, "How can a company that doesn't have such a program compete with a company that does?" Any company hoping to become lean and clean will need its own suggestion system.

Setting Up a Suggestion System

Here are six things to do to set up a suggestion system:

• *Benchmark.* Find out if a nearby company has an effective suggestion system. Join the U.S. Total Employee Involvement Institute (through Productivity Press) and go on one of its tours. Visit a Japanese company with a suggestion system.

• *Buy some books.* Have everyone read *The Idea Book* by the Japan Human Relations Association. It covers the whys and hows.[8]

• *Set up a management-labor idea review committee.* A key reason for the suggestion system is to improve teamwork. Replace the "us versus them" mentality with "us versus the problem." Start with the review committee.

• *Involve everyone.* At Republic a team of janitors designed a system to use cleaning supplies and paper products more efficiently, saving the company about $1000 a month.

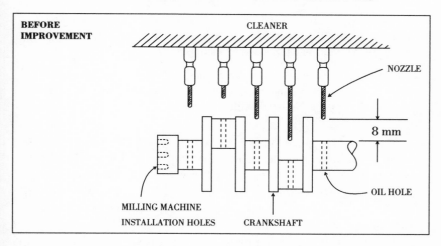

A cleaner discharges air for twenty-five seconds to remove cutting waste from oil holes in a crankshaft. The distance between the tip of the air-blower nozzle and the oil holes in the crankshaft is 8 mm. The cleanliness of the oil holes is just at the tolerance limit.

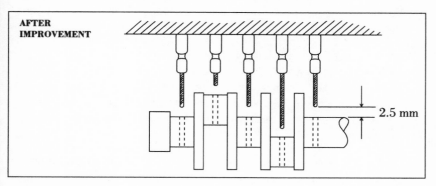

From *The Idea Book: Improvement Through TEI* by the Japan Human Relations Association. English translation copyright © 1988 by Productivity Press, Inc., Portland, OR. Reprinted by permission.

The length of the air-blower nozzle was increased by 5.5 mm so that the distance between the nozzle tip and the oil holes was 2.5 mm. The air blower could then be run for just fifteen seconds, producing cleanliness at the middle of the tolerance range.

Benefit: By reducing the distance between the oil holes and the nozzle tip, the air-blower effect was increased. This improved the cleanliness and reduced air-blowing time. Savings in energy cost amounted to $642 per month.

- *Make it voluntary.* Good ideas cannot be forced; they must come from self-motivated workers who can see that their ideas are not only welcomed but, in fact, needed by management.

- *Give rapid, specific feedback.* At Republic, "Every suggestion gets a response, and if it can't be done, we tell them why," according to Kelly. Pay people for worthwhile suggestions.

Making any change is hard, and an employee suggestion system is no exception. Russell Maier, president and CEO of Republic, said in 1991, "We're going through a major cultural change, and some people are viewing this as who are the winners and who are the losers. We have to accept that, no matter how long we go, there will be people who are unable to make the change."[9]

An employee suggestion system requires tremendous effort and commitment from the supervisors. Managers must see suggestions not as implying poor management by them but rather as strong evidence of their ability to lead. Managers are poor leaders if they take the view "Don't make too many suggestions. If you have the time, you should be working." One Japanese supervisor expressed the necessary attitude:

> I try to implement all improvement suggestions I receive from my subordinates. If a suggestion is difficult to implement, I ask the worker to consider using a different approach. I do this to trigger new ways of thinking. A suggestion system will succeed on its own if the workers believe the supervisors are interested in their suggestions and ideas.[10]

Supervisors have to be trained in the art of stimulating, managing, and evaluating ideas. Line workers, stunted from long years of being ignored, need to be taught a variety of problem-solving skills.

One definition of a problem is a deviation from the norm: an anomaly. An anomaly could be the machine that is always least productive, the pipe that always leaks, or the fuse that always blows. Anomalies arise when what occurs is not what is expected. Anomalies do not become apparent unless one knows what the ideal, expected state is. Here is where a tool such as statistical process control is particularly powerful. It allows one to understand deviations from the norm in a rigorous fashion. For example, if you keep track of your car's miles per gallon every time you fill the tank, and the pattern is 30, 32, 29, 31, 30, 20, 18, then you know you should probably have someone look at the car.

Since workers do most of the observing of production, they can be most effective only when they are trained to understand what they are seeing and are given tools such as statistical analysis. Graphs make data easier to understand. If your employees are not familiar with standard quality control tools such as Pareto charts and cause-and-effect diagrams, they will need training.[11]

Many employees may be worried about their writing skills. Some may even be illiterate. No employee suggestion system will work if employees feel that their writing skills will be judged or that they will be fired if the truth about their poor education is uncovered. A company should help to train people in all communication skills or supply help in transcribing ideas. Consider using a trained facilitator to take suggestions verbally.

Employees should know the cost of the raw material, as well as the cost of parts, tools, energy, and office supplies. *Training* magazine reported that at GM's Saturn plant, "one person checks scrap and receives weekly reports on the amount of waste. If the line on the chart is rising, she reminds everyone during a team meeting that they need to be more careful. Since team members know the cost of each part, they know how much money their scrap costs the company."[12]

Finding the root cause of problems requires even more training. Replacing a fuse responds only to a symptom. Asking *why* five times can track down the source of the problem. Being able to answer each *why* requires extra knowledge. As Harvard Business School Professor David Garvin explains, assembly line workers at most Japanese air-conditioning plants are initially trained for every job on the line, "even though they were eventually assigned to a single work station. . . . Japanese workers were thus more likely to track defects to their source and propose effective remedial action." This training process lasts six to twelve months, whereas comparable American training periods last one to two *days*.[13]

Japanese companies are willing to make such an investment because of the tremendous benefits they see from a suggestion system:

Increased profit
- Improved productivity
- Reduced costs
- Improved safety and reduced accidents
- Improved quality

Improved teamwork
- Improved communication between worker and supervisor
- Improved communication among fellow workers

Enhanced worker ability
- Increased motivation
- Increased problem-solving ability
- Increased cost awareness

For a company committed to improving constantly and becoming a learning organization, a suggestion system is essential. The positive feedback from the suggestion system, coupled with training, nourishes worker creativity and curiosity. "In the beginning, I wrote suggestions based solely on my experience," said one Japanese employee. "However, when I started to run out of things to suggest, I bought books, studied about different materials, machines, and began using this knowledge for greater creativity. I also drew flowcharts on my own to look for waste, inadequacy, or inconsistency."[14] That is how one single employee was energized by his company's search for improvement.

SYSTEMIC PROBLEMS: THE MOST DIFFICULT TO OBSERVE

In the beginning workers will naturally focus on their own operations. Getting workers or anyone else to focus on process changes requires much more training. The easiest problems to spot—anomalies—are deviations from the norm. Systemic problems are hard to spot because they are cases in which the norm itself is the problem; that is, the result the company is expecting is itself flawed.

If, to use the earlier example, you tracked your car's fuel economy as 30, 32, 29, 31, 30, 20, 18, you would probably want to look at the engine. A tune-up may solve that problem. But what if the real problem is that a different engine could be getting twice the fuel economy, for an extra 10 percent initial cost? The more efficient motor might have a ten-year life-cycle cost that is 40 percent lower. Suppose a company has 100 motors, as many factories do, and they are all well maintained, but every one is inefficient.

Seeing such "anomalies" can be much more difficult. According to a 1991 study, "When Do Anomalies Begin?," which followed up the work of Thomas Kuhn,

> An anomaly in science is an observed fact that is difficult to explain in terms of the existing conceptual framework. Anomalies often point to the inadequacy of the current theory and herald a new one. It is argued here that certain scientific anomalies are recognized only after they are given compelling explanations within a conceptual framework. Before this recognition, the peculiar facts are taken as givens or are ignored in the old conceptual framework.[15]

If the scientific community, which is dedicated to observing the world and explaining it, can ignore observations that undermine its worldview, how much more common is this tendency in the business community? In many businesses large inventories may be taken as a given, and high resource use, scrap, and pollution may be ignored entirely. For a company to see these deeper *process* flaws, it needs an entirely new conceptual framework, a new orientation. The primary goal of this book is to provide that new worldview: lean and clean management.

SAVING JOBS THROUGH ENERGY EFFICIENCY

As at Republic Engineered Steels, the management and workers of most companies are typically not open to a new worldview until they are in deep trouble, and change is forced on them. Consider the Sealtest ice cream plant in Framingham, Massachusetts, which was on the brink of extinction in 1989. The Chicago-based conglomerate Kraft General Foods had been closing its oldest and least profitable plants. The Framingham plant, which makes six types of frozen desserts, had been built in the early 1960s, making it one of the oldest—and least efficient—of Kraft's plants. To reduce costs the plant had already gone to a four-day workweek, but more cuts were necessary. Closing the plant would have meant the loss of 200 jobs.[16]

The plant manager, Joseph Crowley, got help financing a sixteen-day audit of the plant's energy needs from the state's Office of Energy Resources. The audit identified $3.6 million in cost-effective efficiency improvements. A senior Kraft executive later told Massachusetts Governor William Weld and former U.S. Secretary of Energy James Watkins that many senior Kraft executives thought the project was impossible. Those executives were stuck in the old paradigm in which such amazing opportunities simply do not exist or cannot be seen. Kraft went forward with the energy upgrades only after the local utility, Boston Edison, offered them a unique deal: Kraft would put up all the money for the project, but would recover that money minus the first year's savings through quarterly incentive payments from the utility; payments would be based on verification of the energy savings for two years. In other words, the utility would guarantee that the project had a one-year payback. Because the $3.6 million project reduced energy use $350,000 a year, Boston Edison paid Kraft $3.25 million. Enough energy was saved to power 1000 homes.

The project made sense for Boston Edison for several reasons. It was more cost-effective for the company to reduce summer peak demand than to build additional generating facilities (see Chapter 5) and possibly finance environmental cleanup of pollution caused by the new power plant. The utility also enhanced the Sealtest plant's ability to remain a customer; it had been the utility's fifty-fifth largest electricity consumer in 1987. And the project provided good publicity for the utility.

The project was even more beneficial to Kraft. It turned an aging plant into a highly competitive state-of-the-art facility that the conglomerate is seeking to duplicate in other parts of the country. A Kraft spokesman noted, "The reduction in energy costs from 7.5 cents to 5.5 cents per gallon of ice cream gives us a major competitive edge."

Workers from all the plant's divisions were included in implementation. The union plant had recently embraced Total Quality

Management, and, Crowley notes, the project was "a perfect example of taking the philosophy into action." The original audit had had very little input from the plant's workers. As a result, the plant's engineers were resistant to the proposed changes, which included replacing the entire refrigeration system. In response, Crowley put together an implementation team that included the plant engineer, the refrigeration engineer, the chief electrician, operating personnel, plant management, and utility representatives. The team decided to modify the existing system, rather than replace it.

The project had a "phenomenal effect on our employees," according to Crowley. It created a feeling that every plant employee had a personal stake in its success. After learning that the efficiency upgrades would include a computerized refrigeration control system, one foreman quietly signed up for a night-school computer course to help him prepare to operate the new system. While some of the improvements were specialized to the needs of an ice cream manufacturer, others—such as using waste heat from the compressors to provide hot water for the plant's daily washdowns—could be used in a variety of industries. That cascading of waste heat not only saved energy, it saved a million gallons of water a year. The project, which won the Association of Energy Engineers Project of the Year Award for 1992, also included a motor and lighting upgrade.

The productivity gains were enormous. The improved refrigeration and air-handling system were much more efficient. The system blew more air and colder air, and it defrosted the air handler faster. As a result, the time required to harden the ice cream was cut in half. *The overall result was a 10 percent across-the-board increase in productivity,* which in the long term will be worth more to the company than the energy savings, says Crowley. In some sense, the productivity gains are far greater, considering that the alternative might have been a shutdown of the entire plant. Today, the plant is expanding production and has added manufacturing jobs.

A team effort—management, workers, the state government, and the local utility—led to process improvement that increased productivity, raised competitiveness, and sustained and increased jobs, while saving electricity and water, reducing air pollution, and even eliminating the use of Freon (a chlorofluorocarbon) in the refrigeration process. One more example of the power of lean and clean.

Although most companies will not consider radical change until they are financially desperate, when it may be too late, very rarely a company will proactively embrace lean and clean. One example is Compaq Computer.

4

A FOCUS ON PREVENTION: THE CASE OF COMPAQ COMPUTER

Rather than building buildings for the sake of the buildings, Compaq was building buildings for people and productivity.
—RON PERKINS, former facilities manager, Compaq Computer

Few companies as a whole can be said to be lean and clean. Most of the case studies in this book are about companies that have performed only one task in a lean and clean way. One company that achieved far more than most far earlier than most is Compaq Computer.

Compaq is a particularly interesting case because it put so many of the principles of lean and clean into profitable use in the 1980s. Compaq succeeded with a focus on prevention, delegation of authority, a common corporate vision, and a faster product cycle than the competition. Compaq demonstrated how resilient a lean and clean company can be in the 1990s.

Compaq Computer is one of the great American success stories. Founded in 1982, it achieved the highest first-year sales in the history of U.S. business, $111 million, with its first product, the Compaq portable. By 1985 it had become the world's second largest manufacturer of personal computers after IBM, and *Forbes* wrote, "Nobody in the computer business has a better marketing strategy than Compaq." In 1987, after a mere five years, Compaq surpassed a billion dollars in sales—faster growth than any other company in history.[1]

Ron Perkins, the facilities manager, explained the company's common orientation: "I attribute Compaq's success to the company's clarity of purpose. There was . . . a clear mission to succeed at 'being a player' and 'doing it better' in the personal computer market." In short, beat IBM.

Compaq took only nine months to go from computer concept into prototype and then production. By contrast, IBM took three years— four times as long. As Compaq got bigger in the late 1980s, it stayed nimble with a systems approach to design and manufacture. It developed products with small teams that included marketers, designers, engineers, and manufacturing experts. Quality circles met regularly, and Compaq used a U-shaped assembly line, to allow workers at the beginning and end to confer.

Perkins used the prevention philosophy of the chief financial officer, John Gribi ("Apply capital dollars to reduce future expense") to justify five- to seven-year life-cycle costing. Perkins notes that to the facilities people, "the concept of borrowing money to pay for a change was completely foreign." But to the treasurer, it made a lot of sense—if the change had a higher return on investment than the interest rate on the borrowed money. As discussed in Chapter 2, Perkins broke down the barrier between the traditional two-year payback mind-set in the facilities division and the treasurer's mental model of a 14 percent return on investment, translating Gribi's vision into the reality of energy efficiency.

Perkins has said, "The role of facilities managers is to embody the corporate culture in brick and mortar." Today Perkins is president of Supersymmetry USA, Inc., an energy-efficiency consulting group in Houston "based on a systems approach with specific expertise in HVAC [heating, ventilation, and air conditioning]." He relates his experience in a case study, "The Compaq Experience: Corporate Dynamics and Energy Efficiency," written in collaboration with Ted Flanigan, an energy expert who runs IRT Environment in Basalt, Colorado, which analyzes the most effective energy-efficiency programs in North America.

At Compaq "every effort was made to create a setting that stimulates creativity and enhances individual effort and productivity," whether it was a heavily wooded landscape or sophisticated manufacturing systems. Designing the work environment to increase productivity requires first asking employees what they want. Compaq repeatedly surveyed and interviewed its workers to find out how they felt about their current workplaces and what they would like to see in a new facility.

The single most common response concerned daylighting. People want as much natural light as possible in their offices and would like to be able to see outside. Perkins "tried to maximize the amount of natural light to people who didn't have outside windows in their offices by putting in sidelights, a window beside each door. This allows a person with an interior office to look outside through the windows of the person across the hall." Other design measures include top-floor skylights, lobby-level atriums, and glass-enclosed walkways. Coffee break areas were moved to the atrium to increase interactions between manufacturing and office workers. People were given a fair amount of individual control over the temperatures in their work spaces. Every exterior office had a thermostat. Clusters of three or four interior offices also had thermostats.

Compaq made no attempt to measure the effect of these and other efforts on productivity, but, as we will see in Chapter 6, similar changes in other companies have led to significant productivity increases. No doubt they contributed to the high growth in productivity and profits that Compaq achieved in the 1980s.

In 1986, Perkins hired Eng Lock Lee of Singapore's Supersymmetry company, to improve Compaq's HVAC systems. Lee took the Gilbreth-Ohno approach of repeatedly asking questions. At his first visit to the Houston-based firm, he asked, "How big is the [cooling] load?" "Why are you using 3000 tons of refrigeration?" "How do you know your peak load?" More often than not, Perkins could not give immediate answers.

Lee also strongly believed in measurement. As Perkins later wrote, "You get what you measure. Measure the performance of systems." Total Quality Management is based on the same philosophy. Measuring costs a little more up front, but the improvement in process that results from knowing what is actually happening more than pays for itself, thereby reducing life-cycle costs.

Lee examined Compaq's central cooling plant, which circulated a "river of water" through a twenty-four-inch pipe to all Compaq's buildings. He concluded that the primary pumps, as well as the secondary pumps feeding each building, were oversized. He recommended turning the secondary pumps off, shutting down eight pumps using a total of 240 horsepower. This simple but "radical" change not only improved overall system performance but saved Compaq $100,000 a year. It took only four hours to do. When Compaq was taking bids for replacement cooling towers, Lee recommended units that were ten times as efficient. The payback was only eight months. The Compaq philosophy was, "Anything less than a year's payback is free."

On a six-month sabbatical, Perkins learned of Rocky Mountain Institute and its work on efficiency, particularly lighting and the systems integration of lighting with HVAC and building design. It was, in his words, "a critical connection." Improved lighting alone in the latest Compaq buildings would have provided savings with a direct 2.7-year payback. But coupling the improved lighting with a better insulated building shell allowed the use of smaller HVAC units, which resulted in zero marginal costs when considering the entire building. In other words, the buildings cost the same as the earlier ones to build but had significantly lower energy costs. Year after year these efficient buildings have saved Compaq hundreds of thousands of dollars.

Typical heating, ventilation, and cooling systems for small buildings are overdesigned by 50 percent, while in larger buildings, overdesign of 100 to 200 percent is common. Even 400 percent has been known to occur. Some overdesign is useful for flexibility and resilience. Perkins has said, "My approach was to build resiliency into the systems. If 1,000 tons of cooling is needed, buy four 300-ton units and run 'em parallel. This provides built-in redundancy in case of failures and allows for better part-load performance."

Such an approach makes sense when energy-efficiency design considerations have been incorporated right from the start. But for the horribly overdesigned HVAC systems of most buildings, the costs are large and pervasive, and not merely because chillers, with all associated equipment, cost about $1500 per ton of cooling.

Oversizing leads to a variety of operating problems that add to this cost.

First, an oversized chiller means operating at part load, and, as with any motor, this means reduced efficiency. At full load, a motor can achieve greater than 90 percent efficiency. At partial load, particularly under 50 percent, efficiency drops off sharply. Second, it is difficult to control a grossly oversized HVAC system operating at partial load. Control systems and valves are imprecise at the low end of the control range. It's like using a chain saw to cut butter. Or trying to pick up a pin wearing gloves: You can do it if they're skintight, but it becomes taxing and time-consuming when they are three sizes too big. Fundamentally, an oversized HVAC is just like excess inventory; it adds cost without value and is a certain sign of rampant waste.

LESSONS LEARNED AT COMPAQ

At Compaq, Ron Perkins learned several lessons that helped make his energy-efficiency work profitable. They include improving observations and monitoring, focusing on the end users and their tasks, and training support staff:

• Gather more information before the design period. Spend money for information. At Compaq this approach led to lower overall costs, totally inclusive of the costs of design, architecture, consultants, and buying the land and building the spaces.

• Make energy-using systems key commodities and integral parts of the whole facility. Teach the value of collecting data and making operational changes as necessary. Monitoring and evaluation allow for greater energy efficiency and smoother running facilities as problems are caught in very early stages. For example, a valve failure is easy to control. Operators can see it coming if they are carefully monitoring their systems, which of course requires sophisticated, but highly worthwhile, instruments.

• Substitute quality for quantity. For example, rather than a uniform blanket of overlighting, look carefully at the tasks that need to be illuminated and suit the lighting to the task. For air conditioning, supply the appropriately sized equipment, because too much capacity creates inefficiency.

- Recognize that one of the biggest problems in accomplishing efficient building operations is the high turnover of facilities managers
and building operators. The employees assigned to operating buildings tend to be treated like very well paid pump managers. They are
not fully trained in the particulars of the systems they are running.
Thus tools at hand (software, hardware, and technologies) are often
not used. Perkins recommends that companies create primers for
new building managers and operators and introduce them with four
days of training in building operations and a tune-up training course
every six months.

The last point is particularly important. Just as Total Quality
Management fails if the line worker is ignored, no enduring gains
in energy efficiency are possible if building managers and maintenance people are ignored. All employees need to be imbued with
the lean and clean philosophy, to be constantly retrained, and to
be listened to for valuable, money-saving advice. Too often it is
assumed that those who manage the building have nothing to contribute to the company, so they rarely feel loyal and tend not to remain. Yet we have seen that the best building managers turn the
company's vision into reality. If building managers and staff are
not brought into the decision-making process when a company decides to become lean and clean, it may win the battle but lose the
war. They may undermine the improvements, either intentionally
or out of ignorance. Or they may simply leave shortly after a major
retrofit, taking with them irreplaceable experience with the idio-
syncrasies of the building and its HVAC systems.

Behavior researchers estimate that much if not most of the performance improvement opportunities for employees in most organizations are to be found in the environment rather than within the
people. The environment is controlled by the building manager. If
he or she is left out of the picture, chances of improving worker
productivity drop dramatically. Give building managers higher pay
and recognition through job titles to properly reward those making
such a contribution to the bottom line.

Another lesson from Compaq: Pay attention to the details. Companies that do the little things right often do the big things
right. What is particularly striking about Compaq is that few com-

panies growing that fast would take notice of the electricity bill, which equaled only 1 percent of the cost of business. And indeed, according to Perkins, Compaq was "more pleased by being a good corporate citizen and providing its workers with comfortable workplaces in which they could be ultimately productive." Yet Perkins understood the powerful leverage of reducing operating costs: "That measly 1 percent cost of business spent on electricity is not so trivial." He points out that since only about 5 percent of sales ends in profits after taxes, halving the energy cost, which represents about 0.5 percent of total operating costs, increases bottom-line profits by the equivalent of a 10 percent increase in sales.

In an all too familiar scenario, the Compaq story doesn't end with the phenomenal successes of the 1980s. With success and size came bureaucracy and complacency. Layers of middle management—eight altogether—permeated the company. The president and one of the founders, Rod Canion, "didn't take the time to pass on the culture," according to Perkins, "to teach [new employees] what had indeed made Compaq such an instant success." He became less and less accessible and said on more than one occasion, "Change could upset the applecart." The company had lost touch with its quick-response, fast-cycle approach. In the beginning there were no purchasing agents, and managers were free to buy whatever they needed. By 1989 purchasing was questioning Perkins's attempt to buy all the state-of-the-art lighting fixtures from one company, even though no one else manufactured them. After that Perkins could no longer point his finger and request special fixtures or whatever else he needed to reduce life-cycle costs.

Most important, Compaq's primary goal of beating the dinosaur IBM now required a paradigm shift. The market changed, but Compaq didn't change its initial strategy, which was matching IBM's prices but with more innovative personal computers. "We created a pricing umbrella that allowed all the clonemakers to thrive," said Gary Stimac, who leads Compaq's systems division. "And when IBM lost touch with the pricing realities of the marketplace, we walked off the cliff with them."

The new competition was from very low-priced computer

clones, particularly Dell, which entered the market in 1990. By 1991, when Compaq announced a $10,000 color portable that required an electrical outlet to run, Dell and AST introduced comparable machines that could run on batteries and cost only $5000.

Sun Tzu had said, "Know thy enemy and know thyself, and in a hundred battles you will be victorious." Compaq did not know the enemy, and it had lost touch with itself. In the third quarter of 1991, it had a $70 million loss, which led to the firing of 1400 employees (about 12 percent of the workforce). The stock fell from its high of nearly $75 to about $26.

After the third-quarter loss in 1991, Compaq's board of directors replaced Canion with Eckhard Pfeiffer, a German-born executive who had been chief operating officer. According to Compaq's manufacturing boss, Gregory Petsch, "Eckhard told all the top managers to go out and break paradigms."

Pfeiffer has worked to cut cycle time. Compaq reduced the number of times the PC processor board was redesigned before production started from fourteen to three. He has delegated decision making, giving departments more responsibility for working out cross-departmental conflicts. The lean side of lean and clean management remains as important as ever. Petsch notes, "You've done well in this business if you can differentiate something for three to six months."

As of this writing, Pfeiffer's efforts are working. Compaq's share of the PC market doubled from 1991 to 1993. Compared with those of 1991, sales and profits for 1992 were 25 and 63 percent higher, respectively. Among the leading computer makers, Compaq had the fastest growth in the United States in 1993. By mid-1994, its stock price was back at $100, before splitting three for one.

Compaq has begun to practice design for the environment, designing its products to reduce their environmental impact when they are in use and designing them to be easy to disassemble and recycle. For instance, Compaq is one of the charter members of EPA's Energy Star Computer program, to reduce the energy consumed by the computers it sells. Compaq also continues to reduce the energy used in its own production operations. In 1992, for example, its Houston operations cut energy use 9 percent. In April

1993, Compaq eliminated chlorofluorocarbon use in all of its manufacturing processes.

Compaq thrives as one of the nation's first truly lean *and* clean companies. The need for the clean side of lean and clean will be increasingly important for all companies. Whoever best uses prevention, energy efficiency, and clean production to lower costs and increase productivity will achieve unique competitive advantage.

5

ENERGY
EFFICIENCY

People do not want electricity or oil . . . but rather comfortable rooms,
light, vehicular motion, food . . . and other real things.

—AMORY LOVINS, *Soft Energy Paths*[1]

Energy efficiency lies at the heart of lean and clean. Energy efficiency will save more money more quickly than almost any other measure to reduce pollution. Most businesses can reduce energy use 50 percent or more, with rapid payback of all costs associated with the improvement. Done correctly, that is to say, done systematically, energy efficiency may be a more cost-effective way to increase productivity and quality than any other potential investment a company can make. The number of companies that have saved jobs and avoided plant closures through energy efficiency grows every day.

It is astonishing, however, that many companies remain unaware of the remarkable gains in energy-efficient technologies and practices. One survey found that small businesses thought energy efficiency required turning down heat or turning off lights, which were not considered acceptable options, "because a cold, underlit store would discourage customers."[2]

Energy efficiency means providing the same or better energy services using less energy; indeed, the latest efficient technology typically provides more pleasing light, more reliable production, and greater comfort and control. In contrast, energy conservation achieves lower energy use by giving up some quality of service. Energy *conservation* is turning down the thermostat and donning a sweater. Energy *efficiency* is insulating a building or using new high-performance windows. Efficiency, not conservation, will be the focus of this chapter.

Energy savings are bottom-line savings: Paring $5000 a month from operating costs goes straight to profitability. Depending on profit margins, that $5000 could be better than a $100,000 increase in monthly sales, which entails increased costs for materials, labor, production, and overhead. Best of all, the energy savings are permanent, whereas a boost in sales might not be.

Finally, efficiency also lets a company do very well while doing much good, reducing emissions of nitrogen oxides, sulfur dioxide, and carbon dioxide—gases that cause smog, acid rain, and global warming.

IMPROVED LIGHTING

More efficient lighting is probably the fastest way to save energy and reduce costs, while increasing quality and productivity. Lighting consumes about 40 percent of the electricity used in commercial buildings, and another 10 percent or more goes to cool the heat generated from lighting. More important, most lighting systems were designed for people writing on horizontal surfaces or using typewriters. Thus, most buildings are both overlit and mislit for computers and word processors. A 1991 Worldwide Office Environment Index Survey found that eyestrain was perceived as a hazard by 47 percent of office workers—a greater hazard than quality of air, radiation from video display terminals, repetitive strain injuries, or hazardous materials.[3] Improving lighting will almost always boost productivity, as the Reno Post Office found in the example discussed in the preface.

The Environmental Protection Agency has a "Green Lights" program to help businesses reduce energy use and air pollution. This voluntary pollution prevention program helps companies obtain the most current information about which energy-efficient lighting technologies work best for a particular application and provides guidance on how upgrades can be financed. According to the EPA, electricity used for lighting can be reduced up to 90 percent, with a 30 to 60 percent rate of return. Many companies are taking advantage of the tremendous savings participation in the EPA program can lead to. For instance, Elkhart General Hospital in Indiana reduced electricity used for lighting by more than 70 percent over 430,000 square feet, for a total annual savings of $102,000. The upgrade cost only $85,000 and prevented emissions of more than 1500 tons of carbon dioxide, 13 tons of sulfur dioxide, and 6 tons of nitrogen oxides. For an investment of $76,000, the St. Francis Hotel in San Francisco reduced lighting electricity 82 percent over 1 million square feet for a total annual savings of $85,000.[4]

Boeing has made many successful Green Lights upgrades. It has reduced lighting electricity use by 25 to 90 percent in several million square feet of its facilities, with a 50 percent return on investment. Boeing's upgrades saved 130 million kilowatt-hours, reducing its contribution to air pollution by 100,000 tons of carbon dioxide, 8000 tons of sulfur dioxide, and 4000 tons of nitrogen oxides—*a year*. Lawrence Friedman, Boeing's conservation manager, estimates that if every company adopted the lighting Boeing is installing, "it would reduce air pollution as much as if one-third of the cars on the road today never left the garage."[5]

Renton, Washington, is called the aircraft capital of the world because Boeing puts out as many as fifty jets a month there, 737s and 757s, in 500,000-square-foot plants. Some of those plants were in the middle of a lighting upgrade when I visited in the spring of 1993. In some cases, half the building had new lighting, half old. The difference between the areas was like day and night: Crystal clear vision, with excellent color rendition on one side; fuzzy, distracting lighting on the other.

With the new lighting, employees had more control, glare was reduced, and the overall appearance of the plant was improved. Friedman says that after the new lighting was put in, the feedback was "almost mind-boggling." One woman, who put rivets in thirty-foot wing supports, said that she had previously been unable to see inside one part she had been working on and had been relying on touch alone. Now, for the first time in twelve years, she could actually see inside the part. Another riveter reported that with the old lighting a rivet head would occasionally break off, fly through the air, hit one of the fluorescent light tubes and possibly break the lamp. The new high-efficiency metal-halide lamps with hard plastic covers don't break when a flying rivet head hits them, an improvement in worker safety.

Improved lighting also resulted in better quality in the shipping and receiving area. The number of packages sent to the wrong destination declined. No wonder factory workers came up to Friedman to ask, "When are you going to upgrade our building?" Steve Cassens, a lighting engineer for Boeing, says that the first thing machinists with new lighting tell him is that they can read the calipers on their lathes and measurement tools much more easily.

One of the shops that was moved to an area with improved lighting has the job of attaching the jets' interior side-wall panel to a stiffening member using numerous fasteners that leave a very small indentation in the panel. The old lighting provided poor contrast and made it difficult to tell if a fastener had been properly attached. The new lamps make it far easier to detect imperfections, far easier to see whether there are indentations—and hence fasteners—everywhere there should be. The new lighting improved by 20 percent the workers' ability to detect imperfections *in the shop*.

The cost savings of catching errors when they happen is enormous. Friedman says that in the aircraft interior most errors that used to slip through "weren't being picked up until installed in the airplane, where [they are] much more expensive to fix"—several times more expensive. Even worse, "some imperfections were found during customer walk-throughs, which is embarrassing." Embarrassing *and* costly, because "the customer says, 'I don't like the way this panel looks,' and then you have to do a special order to match the interior of the customer's plane."

Although it is difficult to calculate the savings from catching errors early, Friedman estimates that those cost savings alone exceed the energy savings.

Boeing's Energy Office was one of the first to get involved in Boeing's vision of Total Quality Management, called World Class Com-

> **Energy-efficient lighting doesn't just save money and reduce pollution, it lets people see better. It reduces defects, increases quality, and boosts productivity.**

petitiveness. Boeing's quality effort is in part a response to the challenge from the European aircraft manufacturer Airbus Industrie and in part a response to a benchmarking trip fifty top executives took, visiting several Japanese high-quality, fast-cycle production plants. The Energy Office's participation is not surprising, given that its work not only cuts costs but improves quality. The office has weekly classes on kaizen, the Japanese philosophy of constant improvement. Friedman himself teaches some of those classes. He says with a smile, "They call us the Energy Office. I call us a think tank. It's our job to give people ideas."

THE ROLE OF UTILITY COMPANIES

Improved lighting has other advantages that speed up payback. Because the new lights put out less waste heat, they reduce air-conditioning load. Because they last longer, they reduce maintenance costs. A 1991 report by the Office of Technology Assessment relates a typical improvement done for a U.S. postal service office, which cut lighting electricity 65 percent (see table on next page).[6] One reason the return on investment for the post office was so high was that the local utility, San Diego Gas & Electric (SDG&E), has a rebate program: It subsidized most of the cost of the improvement.

Why does SDG&E pay people *not* to buy its product? Because paying people to install energy-efficient equipment is cheaper and less polluting than building new power plants. Power plants take years to construct and commit a utility to increased power generation capacity for decades. Making that commitment requires long-term forecasting of demand, which is usually extremely inaccurate in a world where energy prices, federal and state clean air regulations, and customer environmental concerns are constantly changing.

RETROFIT LIGHTING ANALYSIS

Lighting savings (annual)	$4590
Reduced A/C savings	689
Reduced maintenance	333
Total savings	**$5612**
Retrofit cost	$6855
Utility-paid incentive	5331
Customer cost	**$1524**
Annual return on investment	368%
Payback period	3 months

Additional benefits
- Hedge against future rate increases (possible increased future savings)
- Brand-new lighting system—reduced future maintenance cost
- More pleasing light—less glare
- Additional profits for your business, keeping it more competitive
- Increased marketability of the building

Efficiency is an end-use/least-cost approach. The utility gives its customers better lighting and more comfortable heating and cooling while reducing electricity use. By thinking of itself as a "service company," the utility can better manage demand, which is why this approach is called demand-side management. Generally such an approach requires an enlightened state utility commission, such as California's. Utilities have traditionally been allowed to make a profit from selling more power and building new power plants, but wiser regulations allow utilities to make a profit from investing in whatever new source of energy—including energy efficiency—has the lowest cost. Efficiency lowers risk, reduces capital costs, lowers customers' energy bills, and, with new regulations, increases profits. It also allows a utility to respond more quickly to a changing marketplace.

San Diego Gas & Electric once was California's highest-cost producer of energy; it is now the state's most efficient provider. In 1993 *Forbes* called SDG&E "one of America's 21 most nimble companies."[7] The company philosophy is proactive. "I want to set the agenda, rather than just react to someone else's agenda," says Chief

Executive Thomas Page. "There should be a reward for being smart, rather than just investing money." And because it promotes air pollution prevention, SDG&E is truly a lean and clean company.

When planning to retrofit an office with new equipment, first call your local utility. Most utilities in the country now offer some sort of program to promote energy efficiency. If they offer technical assistance and a rebate, it will speed implementation and payback. *Next call the EPA,* to join its "Green Lights" program. The EPA is a top source of useful information and of a seal of approval that testifies that a company reduced pollution. The Green Lights hotline number is 202-775-6650. *Step three is to find a reputable energy-service company,* perhaps one recommended by the EPA. It will not only assist you in choosing the equipment best for your needs, but may actually help finance the upgrade, in some cases putting up all the capital and taking payment from the energy savings.

BASIC HARDWARE IN A LIGHTING UPGRADE

The basic hardware changes needed for an upgrade to more energy-efficient lighting include

- Reduce the number of existing fixtures (most offices are overlit)

- Replace remaining 4-foot fluorescent lighting fixtures with
  1. more efficient lamps
  2. more efficient control ballasts and optical reflectors

- Replace incandescent bulbs with compact fluorescent bulbs

- Install sensors to turn off lights when room is vacant and photocells to adjust light to changing outside conditions

- Install task lights tailored for the work in each area

The ballast is the transformer of the fluorescent light, providing the necessary current and voltage to run it. The new electronic ballasts are much more energy efficient than the electromagnetic ballasts they replace. Also, they operate the lights at a much higher frequency, which eliminates the flicker and hum that give so many people headaches and eyestrain. The color rendition of the best new lamps is much better, far closer to sunlight than that of the old bluish lamps, which give everything a sickly pale tinge.

Compact fluorescent bulbs last considerably longer than incandescents, and the savings in replacement labor more than compensates for their higher cost. That change alone is profitable, without even counting the savings in electricity. A 1990 article in *Scientific American* written jointly by Amory Lovins and the Electric Power Research Institute noted: "This is not a free lunch; it is a lunch you are paid to eat."[8]

Occupancy sensors are designed to turn off the lights in an unoccupied room automatically. The problem is that early versions of the sensors would turn off the lights even if the occupant remained fairly motionless for a few minutes. Today there are heat-sensitive occupancy sensors, which eliminate that problem.

More complete use of a resource is always important. Increased use of natural sunlight (daylighting) can dramatically reduce lighting electricity in areas near windows, while offering better views and the feeling of more space. If possible, use top-silvered blinds; light-colored finishes, furnishings, and fixtures; and glass partitions atop office partitions. Reliable, efficient lamp controls and photocells for dimming fluorescent lamps are now available. They should be considered for all offices and lighting where daylight is available.

OFFICE RETROFIT

A large New England office area replaced 7023 lighting fixtures with energy-efficient fixtures at a cost of $863,475. Lighting energy was reduced 63 percent, and annual electrical savings were $183,341. The utility rebate was $405,688, so the net cost was $457,787: a payback of 2.5 years. Glare on VDT screens was eliminated. Employees love the new system.[9]

FOCUS ON THE END USER

Never lose sight of the basic systems principle: Focus on the end user. In this case, focusing on the end user of the lighting will open the biggest opportunities to increase savings and productivity.

It is easy to become caught up in the excitement of installing the latest energy-efficient lighting and miss what may well be the main benefit of the changes. One major manufacturer I visited had recently installed the latest fluorescent lamps and electronic ballasts in a 15,000-square-foot facility where the workers do a great deal of detail work, including wiring. The company cut the electricity used for lighting in the facility by 50 percent with a payback of about two years.

But as I walked through the area with the energy manager, it quickly became clear that the most cost-effective improvements, the end-use ones, had been ignored. On one side of the area, about three-fourths of the lights were on even though not a single person was working there. On the other side, all the lights were on, even though only two supervisors were working in one corner. In other words, hundreds of lights, albeit energy-efficient ones, were providing illumination for two workers.

The supervisors explained that, because of the recession, the company had recently gone from two shifts to one. They assured us, however, that the lights were on a timer set to turn off at 5:30 P.M. Since it was nearly 6:00 P.M., we were not reassured. Those lights may well have been on twenty-four hours a day, and no one in the company would have known that money was burning up for no reason. The supervisors also pointed out that their half of the area was controlled by only three switches, so they could have only turned off two-thirds of the lights in any case.

They didn't actually measure the rate of errors, defects, or rework, so they could not give a reliable or quantitative answer to the question of whether the new lighting had reduced errors or defects. They didn't think quality had risen, however, because they did mainly detail work, and the replacement lights were exactly where the old ones had been, about six feet above the work area—too high to be of much help. Many of the workers bought small desk or task lamps at Wal-Mart for six dollars.

Instead of a one-size-fits-all lighting upgrade, the company should have taken the end-use approach. It should have interviewed its workers to find out what kind of illumination they needed. It should then have mocked up several areas with different combinations of overhead lighting and task lamps and allowed workers to pick the lighting they liked best.

The approach the company took used high-efficiency lamps and ballasts to reduce lighting electricity, but it did not record an improvement in quality of work or productivity. The end-use approach would have added the following:

- Task lights tailored for the work in each area

- More individual light switches, to allow workers direct control of the overhead lighting above their work areas

- Fewer overhead lights

- Occupancy sensors to turn the lights off when no one was there

These improvements would have cut lighting electricity use by 80 to 95 percent, depending on whether the lights were on just for the second shift (i.e., sixteen hours a day) or, in fact, all night. Most of these changes are inexpensive, and some, such as reducing the number of overheads, are virtually cost free.[10] The payback period would have dropped to under one year—a return on investment exceeding 100 percent. Moreover, quality would certainly have increased with the improved lighting from the task lamps. Workers would have had lamps they themselves had chosen, and they wouldn't have had to waste their own money buying desk lights. Productivity would no doubt have improved measurably (if, that is, the company started measuring it), and the return on investment would probably have doubled or tripled. The next chapter discusses how one utility took the end-use approach and achieved a return on investment in excess of 500 percent.

Another business, one of the most successful service companies in the country, did a similar one-size-fits-all upgrade with new bulbs and ballasts, cutting lighting electricity by over one-third with rapid payback. Unfortunately, the company never bothered to ask its workers what troubled them most about the old lighting. The answer would have been glare, since every office worker uses a computer. The new lighting also produces glare, and many of the workers simply shut the lights off and either work in the dark or buy themselves desk lamps or special task lamps that attach to a computer. An end-use approach would have used antiglare fixtures, fewer overhead lights, and new task lamps. It would have cut electricity use even more with faster payback. Again, productivity would probably have risen.

The lesson is clear: *The end use approach is invariably the least-cost approach.* Don't do a "one-size-fits-all" lighting retrofit. Don't adopt the checklist mentality: "We did fluorescent lamps with electronic ballasts, so we've done energy-efficient lighting. Now we can move on to something else."

Talk to employees. See what their needs are. They may prefer pale blue fluorescent lamps, which give sharper contrast to the black-and-white drawings of modern blueprints; high-efficiency versions of those lamps are available now too. The maintenance people who clean and replace lamps should also understand the changes, so they don't replace a burnt-out efficient bulb with some inefficient bulb in storage. Get rid of those old bulbs. In fact, better maintenance schedules can also be a source of energy savings. Here's what one engineer recommends:

> Review cleaning crew schedules. In one commercial office building, the building manager was able to save $17,000 per year at no cost whatsoever—he merely changed the cleaning crew schedule to overlap normal business hours to avoid over half the building's lights being left on between 5:00 P.M. and when the cleaning crew normally started.[11]

Set up a trial area first, as Hyde Tools did. Hyde is a manufacturer of cutting blades with 300 employees and an active total quality environmental management program. Doug DeVries, purchasing manager from 1972 to 1992, notes that no amount of money saved will compensate for dissatisfied operators. He did a lighting upgrade from old fluorescents to new high-pressure sodium-vapor and metal-halide fixtures that cost $98,000 (including labor), with $48,000 covered by the local utility. He estimated that annual energy savings would also come to $48,000 (for a payback of about one year), but he still insisted on trying the upgrade in only one area to start. He left in the original fixtures in case workers wanted to change back after an agreed-upon six-month trial period.[12]

"For the first three weeks, a lot of people complained because the new lights cast an orange hue," said DeVries. "But when we experimented by turning the old fluorescent lights back on after six months, there was a near riot of disapproval." Why? For one thing, the new lights made it possible to see tiny specks of dirt on the equipment that holds the blades while they're being worked on. That dirt creates tiny indentations on a blade, called mud holes. The mud holes make the blade defective or difficult to plate, which can lead a customer to reject it.

With the new lighting, DeVries says, "the quality of work improved significantly because we could see things we couldn't see before." DeVries estimates that *the improved quality was worth another $25,000 a year to the company.* Those bottom-line savings are critical to a small company. DeVries notes that every dollar saved on the shop floor is worth ten dollars in direct sales. In other words, the improved quality from the efficient lighting was the equivalent of a $250,000 increase in sales. (Interestingly, DeVries's upgrade was not the first time the company used lighting to improve quality. Isaac P. Hyde, who founded the company in 1875, built his first shop in the form of an H so that all of his workers could examine the finished edges in natural light.)

In the mid-1980s, Control Data's Operations Group in Sunnyvale, California, analyzed its lighting and found that it not only used energy inefficiently but lowered work performance because of glare on keyboards and screens. With the help of an energy management engineer, the company installed high-efficiency bulbs and ballasts, removed unnecessary lights, and improved lighting control at a cost of $15,000. Energy use dropped 65 percent, saving $7000 a year. Reduced lamp replacement costs and labor saved another $250 a year. These savings would have paid for the upgrade in about two years, a 50 percent return on investment. The reduced glare improved the vision and performance of the workers, lowering the number of input errors and raising the group's productivity by an estimated 6 percent, which was worth $28,000 a year to the company. This lean and clean lighting improvement paid for itself in well under a year.[13]

IMPROVED OFFICE EQUIPMENT

Computers and related equipment can consume up to half the electricity in a modern office building. Much of that electricity goes to air conditioning to cool the excess heat produced by the equipment. Rocky Mountain Institute has estimated that up to 70 percent of the energy consumed by office equipment could be reduced in the short term. In the longer term, with new equipment and office redesign, energy savings may exceed 90 percent.[14]

Up to 70 percent of office equipment is left on all the time, even though it is used infrequently. Turning off computers and other equipment when they are not needed does not hurt the hard drive or other components and will save considerable energy. A midsized copier might directly consume $350 worth of electricity a year if left on twenty-four hours a day, a personal computer about one-third that amount.

Get specifications from several dealers before buying a new copier. Many copiers have standby features that save a great deal of energy, yet warm up in a matter of seconds. Consider buying more compact tabletop copiers for small jobs. They use far less energy and can warm up very quickly. Similarly, save the laser printer for important work and use ink-jet printers for most work. Ink jets are far cheaper and use up to 99 percent less energy when printing.

The EPA's Energy Star computers program certifies computers and related equipment as energy efficient. The EPA has worked with a variety of major manufacturers—including Apple, Compaq, DEC, Hewlett Packard, IBM, NCR, Smith Corona, and Zenith—to develop computers that run on less energy and either go on standby or shut off when not being used. Hewlett Packard has a 1993 model laser printer that draws only five watts of power when not being used yet immediately warms up when needed for printing. Fred Forsyth, general manager of Apple's Macintosh division, explained why his company signed up to build and market energy-efficient printers: "Apple wants to minimize the environmental impacts of our products throughout their life cycles."[15]

Finally, consider laptop computers. They use up to 99 percent less energy than desktops, and the best new ones consume almost no standby power. Laptops often have a higher initial cost but a much lower life-cycle cost. Rocky Mountain Institute's Rick Heede and Amory Lovins have found that "the extra cost of the laptop machine can often be repaid by just its energy savings, especially in new buildings." When its energy and building air-conditioning requirements are compared with those of an IBM personal computer, a single laptop can save as much as $950 in direct costs. Operated 2500 hours a year, that one laptop can reduce carbon dioxide emissions by about one ton per year.

Laptops are clean *and* lean. They increase a company's flexibility, allowing more employees to work at home. This telecommuting saves even more energy—and the workers' time—by avoiding commuting. As one example, AT&T has been considering giving many of its salespeople laptops and basing them at home, not only saving energy but reducing office-leasing costs. As computers get even smaller, with improved flat-panel displays and built-in energy management systems, potential energy savings will grow and grow.

The overall systems benefit to your office of energy-efficient equipment is considerable. As Heede and Lovins note, reducing the electricity load can avoid "costly expansions" of air conditioning and ventilation. In new buildings, when coupled with improved lighting and windows, efficient equipment "may even eliminate major mechanical systems altogether."

IMPROVED HEATING AND COOLING

Like most efforts to reduce waste, saving the energy used in heating, ventilation, and air conditioning (HVAC) requires a systematic approach. Do not upgrade HVAC until your lighting and office equipment have been upgraded because so much of the "load" the HVAC system must cool is the excess heat from inefficient lights, computers, and the like.

Similarly, poorly insulated windows are responsible for about 25 percent of all heating and cooling requirements. Such windows allow excess solar radiation into a building during the summer and lose heat during the winter. High-efficiency windows let in the visible light but block the infrared radiation (heat), which avoids most of that loss cost-effectively. Advanced window films that let in the light but block the heat are now available. These "retrofit films" can turn normal windows into superwindows. Many utilities offer rebates for such films.

A thorough analysis of HVAC use can increase efficiency. When Global Turnkey Systems, a small computer system customizer, analyzed its HVAC system, it found that wintertime blasts from office space heaters often drove corridor thermostats higher, stopping the flow of heat to other offices, whose occupants would then add space heaters. Similarly, in the summertime electric heaters were used to neutralize the cold spots caused by the air conditioning. Reengineered vents produced uniform temperatures. Now there are few space heaters around. The utility bill was reduced $2000 a month.[16]

New control systems take advantage of a building's tendency to store heat and change temperature slowly, saving 30 percent to 70 percent of air-conditioning costs. In the summer, in the early afternoon, it is more efficient to shut off the air conditioner and allow the inside temperature to rise slowly until it exceeds the comfort range shortly before the building is no longer occupied. At night electricity is often cheaper, and the outside air is cooler; the AC can be turned on to precool the building, which reduces AC use during the day, when rates are high. Comfort is not compromised since the building's coolness delays heat buildup (which is even more true if upgraded lighting and windows have already reduced the heating load).

Finally, improved maintenance of HVAC systems and of the building shell can save a considerable amount of money. As Ken Teeters, plant engineering manager for Harrah's Hotel and Casino in Las Vegas, says, "You have to identify a critical energy management flaw that can happen at any operation: . . . failure to properly maintain existing equipment. It can result in as much as a 30 percent energy waste."[17] Very simple maintenance jobs can reduce your energy bill at very low cost:

- Check thermostats. See to it that they give true readings.

- Replace all air filters regularly.

- Inspect and clean duct work. Repair leaks large and small.

- Seal masonry and concrete cracks with caulking or other materials.

- Check weather stripping. Replace when necessary.

- Have an expert clean and adjust the boiler and air conditioner, and check the refrigerant for leaks.

A SYSTEMS APPROACH: THE COMSTOCK BUILDING

Once a company has improved lighting, office equipment, insulation, and windows, *then* it can install a new energy-efficient HVAC system, which might well be twice as efficient as the existing system. A high-efficiency HVAC might cost a little more, but earlier measures will have reduced the heating and cooling load, allowing use of a smaller HVAC system. The savings from the smaller HVAC system will at least pay for the system—and may pay for the initial investments in improved lighting, insulation, and windows.

Since retrofitting is highly profitable, designing a building efficiently from the start must be superprofitable. The ten-story, 175,000-square-foot Comstock building, completed in 1983, cost $500,000 less to build and has half the normal operating cost of other large Pittsburgh office buildings. The architects, Burt Hill Kosar Rittelman Associates, won the 1986 ASHRAE Energy Award for new commercial building for their innovative design.[18]

How were the energy savings achieved? The architects used high-efficiency lighting, an energy management system, and efficient heat pumps to provide heating and cooling. Good insulation, plus careful placement and design of windows, allowed the use of a smaller heating, ventilation, and cooling system than would typically be needed. The heat pump system cost half as much as a conventional system. Net savings, even with the more costly windows, exceeded $500,000. The Comstock building achieves high efficiency even though a very high lighting level was used in the 30 percent of the building that housed an engineering firm's drafting department. This reduced energy use is also remarkable because the air flow rate was two to four times higher than that used in most buildings a decade ago (although it is the recommended standard today, to make the air as healthful as possible).

The Comstock building was also built on a very tight schedule: eighteen months for design and construction. How did architects achieve a design that was both fast-cycle and green? First, the building owner was "open to doing things differently," in the words of Paul Scanlon, director of engineering at Burt Hill and the engineering project manager on Comstock. The "very enlightened owner" gave the architects the freedom to experiment. Second, "We did a lot of life-cycle analyses," says Scanlon. The life-cycle analyses did not just cover operations and maintenance, as is often the case, but also included construction costs and leasing income. Only such comprehensive analyses can lead to overall cost minimization.

Finally, and most important, the architects used a cross-functional team with "an unusually high degree of coordination among all members of the design team," as ASHRAE noted in its award. The functional disciplines on the team included architects; engineers for HVAC, electrical, communications, and structural systems; space planners; and interior designers. Scanlon himself had graduated from a special five-year program that included both architectural and engineering courses, so he was particularly suited to manage interdisciplinary work.

The fast pace drove the interdisciplinary process. "We had such a tight time schedule that there was no other way to go," says Scanlon. "Representatives from each discipline locked themselves in a room from day one." Thus, a systematic, integrated approach to building design can lower both first costs and life-cycle costs, while reducing pollution and meeting a very tight schedule. This is lean and clean building design.

If the benefits of getting the design right the first time are so significant, why isn't every new building lean and clean? First, there is a general lack of awareness of the opportunities among developers, architects, and engineers. As architect William McDonough says, most American buildings of the past few decades are "monuments to the designer's ignorance of where the sun is." Proper choice of architectural form, envelope, and orientation by themselves can often save one-third—44 percent in one recent California design—of the building's energy at no extra cost.[19]

The barriers to lean and clean design, however, go much deeper. As one example, Scanlon notes that "we put in a lot of work to save the owner $500,000, which cut into our design fee." Architects are typically paid a percentage of the construction cost. If they do energy planning in a traditional manner, they don't have to spend much of their time on it, and the fee may actually be higher. In Scanlon's words, "It's a lot easier to put out a standard system. You don't do a lot of research. You don't interact with the client." In other words, the incentive system is really a disincentive system.[20]

Another serious problem is that the developer, architects, and engineers are not the ones who will be using most buildings. They rarely talk to the people who will be using a building—the employees of whatever company leases space in the building—to find out what their needs are. They hardly ever perform customer surveys after the building is occupied to find out what employees did or did not like. But without talking to customers, how can a good product be built?

Many smaller companies rent space from a real estate company. The landlord pays the energy bill and passes it on, prorated by floor space and often with a small percentage increase tacked on for profit. The renter has no incentive to reduce energy costs, and the landlord may well have an incentive to keep costs high. For renters whose floor space is metered, here is one possible solution: Offer to pay the landlord a fixed amount based on last year's energy bill (adjusted for inflation and changing energy costs) and offer to upgrade the lighting (and other equipment) under the condition that any savings go to you. This option would require that the rented floor space be separately metered to measure the effect of the efficiency upgrades. Of course, managers primarily concerned with gains in productivity and quality may want to upgrade their offices whether or not they will obtain any of the energy savings.

Failing to use lean and clean design—a true customer-driven end-use approach—is missing an enormous opportunity. As the next chapter clearly shows, lean and clean building and office design can have benefits that far exceed the energy savings: improved productivity, better quality work, lower absenteeism, and higher employee morale.

6

LEAN AND CLEAN DESIGN
FOR OFFICE PRODUCTIVITY

There is a common chord in all this that will be heard; and it is not a plea for ugliness. It is a plea for first principles—for less heat and parasitism, and more light and pragmatic integrity; for less architecture in quotation marks and more engineering. I feel that the sceptre has all but passed from the hands of the architect to the hands of the engineer, and if it is ever to be the architect's again, he must take it from the engineer by force of superior virtue.

—FRANK LLOYD WRIGHT, 1909, replying to a critic's evaluation of the daylighted Larkin Building[1]

A systems approach to office and building design is a pillar of lean and clean management. Designing work spaces for the end user—the worker—does not merely save money and save energy (thereby reducing air pollution). Lean and clean office design can also increase worker productivity and work quality, thereby dramatically reducing the life-cycle cost of an office or building.

In case after case the savings from the jump in productivity overshadow the energy savings. When the productivity and quality gains are counted, lean and clean designs that would have paid for themselves in energy savings in a few years suddenly pay for themselves in several weeks. In the early 1980s, for example, Pennsylvania Power & Lighting Company had been increasingly concerned about its lighting system, especially in a 12,775-square-foot room for its own drafting engineers. According to Allen Russell, superintendent of the office complex, "The single most serious problem was veiling reflections, a form of indirect glare that occurs when light from a source bounces off the task surface and into a worker's eyes." Veiling reflections wash out the contrast between the foreground and background of a task surface—such as the lines on a drawing and the film on which they're drawn—making it much more difficult to see. This in turn increases the time required to perform a task and the number of errors likely to be made. Russell adds, "Low quality seeing conditions were also causing morale problems among employees. In addition to the veiling reflections, workers were experiencing eye strain and headaches that resulted in sick leave."[2]

After considering many suggestions, the utility decided to upgrade the lighting in a 2,275-square-foot area with high-efficiency bulbs and ballasts. Perhaps more important, while the old fixtures ran perpendicular (north-south) to the workstations (east-west), the new fixtures were installed parallel to reduce veiling reflections. To improve lighting quality still further, the fixtures were fitted with eight-cell parabolic louvers—metal grids that help reduce glare. Russell notes,

> Generally speaking, it can be said that we converted from general lighting to task lighting. As a result, more of the light is directed specifically to work areas and less is applied to circulation areas, creating more variance in lighting levels which upgrades the appearance of the space.

With veiling reflections reduced, less light was needed to provide better seeing conditions. Russell believes this is a general principle: "As lighting quality is improved, lighting quantity often can be reduced, resulting in more task visibility and less energy consumption."

Finally, local controls were installed, according to Russell, "to permit more selective use of lighting during cleanup and occasional overtime hours." Previously, all the lighting was controlled by one switch, and every fixture had to be on during cleanup. With multiple circuits, maintenance crews can now turn the lights on and off as they move from one area to the next.

Russell performed a detailed cost analysis, comparing the initial capital and labor costs of purchasing and installing the new lighting with the total annual operating costs, including energy consumption, replacement lamps and ballasts, and fixture cleaning and lamp replacement labor. He found that the total net cost of the changes amounted to $8362; lighting energy use dropped 69 percent. Total annual operating costs fell 73 percent, from $2800 to $765. The $2035-a-year savings would have paid for the improvement in 4.1 years, a 24 percent return on investment.

Under the improved lighting, however, productivity jumped 13.2 percent. In the prior year it had taken a drafter 6.93 hours to complete one drawing—a productivity rate of 0.144 drawings per hour. After the upgrade, "the time required to produce a drawing dropped to an average of 6.15 hours, boosting the productivity rate to 0.163 drawings per hour." The productivity gain was worth $42,240 a year.

The annual savings derived from the lighting system upgrade increased from $2035 to $44,275, with energy savings, maintenance savings, and productivity improvement benefits having been specifically documented. Simple payback dropped from 4.1 years to 69 days. *The productivity gain turned a 24 percent return on investment into a 540 percent return.*

"Not only is this an amazing benefit," comments Russell, but "it is only one of several." Before the upgrade drafters in the area had used about seventy-two hours of sick leave a year. After the upgrade the rate dropped 25 percent to fifty-four hours a year. "Improved employee morale also is noticeable." The better appearance of the space, reduced eye fatigue and headaches, and the overall improvement in working conditions all helped boost morale. Finally, supervisors reported that the new lighting has reduced the number of errors. Better lighting means better quality work. Russell says of the reduced error rate, "We are unable to gather any meaningful data on the value of these savings because any given error could result in a needless expense of thousands of dollars. Personally, I would have no qualms in indicating that the value of reduced errors is at least $50,000 per year." If this estimate were included in the calculation, the return on investment would exceed 1000 percent. The new lighting slashed operating costs, reduced energy use and pollution, raised productivity, reduced sick leave, boosted morale, and cut errors. Drafters did higher quality work in less time. They became faster cycle. The lighting improvement was truly lean and clean.

The Superior Die Set Corporation of Oak Creek, Wisconsin, found a similar result. The company's complete lighting upgrade, which included more efficient lighting and occupancy sensors, cost just under $3000. The reduced energy and maintenance costs came to $1750, a payback of twenty months. One of the efficiency upgrades was aimed at reducing veiling reflections in the drafting area. The improvements cut the time it took a drafter to turn out one drawing to 6.3, hours down from 7.1 hours, an 11.3 percent reduction. The productivity gain in the drafting area was worth $37,500. At the same time the upgrade significantly reduced glare in the computer operations area; subsequently computer operations downtime fell by four hours per month, saving another $5700. The total savings to the company came to about $45,000 per year. The productivity gains reduced the payback from twenty months to less than twenty-four days—another lean and clean improvement.[3]

Designing offices and buildings to reduce energy use boosts productivity. The overall results—presented here for the first time—will not particularly surprise readers. Lean and clean thinking forces designers to adopt the end-use/least-cost approach. They must consider how to give employees the lighting and the comfort that they want, which usually means giving them more control over lighting, heating, and cooling. Top managers have always been rewarded with corner offices and thermostats—daylight and control. With lean and clean design, so is everyone else.

Green office design invariably requires a systems-oriented focus on the workplace environment. It requires "more engineering," as Frank Lloyd Wright pleaded for eight decades ago. Designers come to see the workplace as a "production system," in the words of architect Lee Windheim, who designed the Reno Post Office upgrade as well as the heavily daylit Lockheed building that raised productivity 15 percent. Systematic design and more worker control over the work environment are all techniques that Shigeo Shingo and Taiichi Ohno used to boost productivity in manufacturing. No wonder these measures boost productivity in the office as well.

Work done at Western Electric's Hawthorne Works in the 1920s and 1930s suggests that contrived experiments to monitor the effect a workplace change has on productivity can be complicated by the special conditions of the experiment, particularly the interaction between the worker and the experimenter. Indeed, some have come to see the Hawthorne effect as implying that changes in the physical environment have an effect on worker performance *only* because those changes signal to the worker the interest and concern of management. In fact, many of the Hawthorne researchers themselves did not believe that productivity rose simply because workplace changes signaled special management interest. For them the experiments showed that productivity can be enhanced by a more cooperative relationship between management and labor, a greater identification by workers with the goals of management, and more effort by management to treat workers with respect and to be responsive to their needs and abilities.

Perhaps more important, though, subsequent analyses have called into question the experimental methods and results of the Hawthorne research, and a major 1984 study of the effect of office design on productivity found direct correlation between specific changes in the physical environment and worker productivity. In any case, the real-life cases described in this book bear no resemblance to the contrived experiments done by the Hawthorne researchers.[4]

The total life-cycle costs of an office building over thirty years break down roughly as follows:[5]

Thirty-Year Life-cycle Costs of a Building	
Initial cost (including land and construction)	2%
Operation and maintenance	6%
Personnel costs	92%

In a typical building, energy costs average $1.50 to $2.50 per square foot, while salaries average $150 to $250 per square foot. That is why the productivity savings of lean and clean design dwarf the energy savings. Cutting energy use in half might save $1.00 a square foot; boosting productivity 5 percent will save $5.00 a square foot. Failing to design a building to increase worker productivity means ignoring 92 percent of its life-cycle cost. Yet most designers do just that. Let's look at some case studies of designers who did better.

THE NMB BANK BUILDING

In the late 1970s the directors of one of Holland's largest banks, the Nederlandsche Middenstandsbank (NMB Bank), decided to build their new headquarters using green design techniques, including state-of-the-art energy efficiency. Construction began in 1983 and finished in 1987. The 538,000-square-foot office building houses 2400 workers in "10 slanting towers arranged in an irregular S-curve with gardens and courtyards interspersed," as described by Bill Browning, director of Rocky Mountain Institute's Green Development Program.[6]

The interior has many plants and gurgling streams of water called flow-form sculptures. The building makes extensive use of daylighting. One design criterion is that no desk can be more than twenty-three feet from a window. Most of the light comes from the sun, and the rest is provided by task lighting, decorative wall sconces, and a few overhead fixtures.

The NMB building is designed to store the heat from passive solar measures, as well as from the lighting, office equipment, and people. In addition, an on-site cogeneration system (see Chapter 8) provides hot water. The bank has no air conditioning but relies on mechanical ventilation, natural ventilation through operable windows, and a backup absorption cooling system powered by waste heat from the cogeneration system.

An adjacent bank built at the same time for about the same cost uses five times as much energy per square foot. The energy-efficient design requirements for the NMB building increased construction costs by about $700,000—but energy savings come to about $2.4 million a year, which paid for the efficiency measures in three months. Absenteeism has dropped 15 percent, which saves the company more than $1.0 million a year.

The head of NMB's real estate development subsidiary, Tie Liebe, attributes the drop in absenteeism to the improved work environment. Liebe also believes the building has vastly improved the bank's image: "NMB is now seen as a progressive, creative bank, and its business had grown dramatically." In 1978 NMB was the fourth largest bank in the Netherlands. Today it is number two.

Not surprisingly, the design team for this lean and clean building was cross-functional, including an architect, a structural engineer, an energy expert, acoustics advisers, interior designers, a landscape architect, artists, and a representative appointed by the bank. In addition, employees were involved in design and technical considerations. For example, workers wanted operable windows rather than air conditioning. They also wanted "natural" material wherever possible, to minimize polluting materials such as foamed plastics made with CFCs or those that might emit noxious gases such as formaldehyde.

LOCKHEED BUILDING 157

Most of the energy savings and productivity gains I have discussed here come from more efficient use of energy, particularly lighting. But the best, cheapest, cleanest, and most desirable form of light is completely renewable. It comes from the sun.

One of the most successful examples of daylighting in a large commercial office building is Lockheed's Building 157 in Sunnyvale, California. In 1979 Lockheed Missiles and Space Company commissioned Leo J. Daly (the firm that designed the Reno Post Office, discussed in the preface) to design a new 600,000-square-foot office building for 2700 engineers and support people.[7]

The architects posed a question to Lockheed: "If we could design a building for you that would use half as much energy as the one you're planning to build, would you be interested?" Lockheed said yes, and Daly's architects responded with a design for energy-conscious daylighting that was completed in 1983 for $50 million.

Daly used fifteen-foot-high window walls with sloped ceilings to bring daylight deep into the building. "High windows were the secret to deep daylighting success," says the project architect, Lee Windheim. "The sloped ceiling directs additional daylight to the center of each floor and decreases the perception of crowded space in a very densely populated building." Daylighting is also enhanced by a central atrium, or lite-trium, as the architects call it. The lite-trium runs top to bottom and has a glazed roof. Workers love it. They consider it the building's most attractive feature.

Other light-enhancing features include exterior "light shelves" on the south facade. These "operate as sunshades as well as reflectors for bouncing light onto the interior ceiling from the high summer sun," in the words of two researchers from Lawrence Berkeley Laboratory. "In the winter, the interior light shelves diffuse reflected light and reduce glare during lower winter sun angles." The overall design "separates ambient and task lighting, with daylight supplying most of the ambient lighting and task lighting fixtures supplementing each workstation." Finally, continuously dimmable fluorescents with photocells were installed to maintain a constant level of light automatically and save even more energy.

The daylighting has saved Lockheed about 75 percent on its lighting bill. Since daylight generates less heat than office lights, the peak air-conditioning load is also reduced. Overall the building runs with about half the energy costs of a typical building constructed at that time. Although Daly's energy-efficient improvements added roughly $2 million to the cost of the building, the energy savings alone were worth nearly $500,000 a year. The improvements paid for themselves in a little over four years.

But the daylighting was part of a larger plan to boost worker productivity. The open office layout and a large cafeteria were designed to foster interaction among the engineers. At the same time workstations were tailored for employee needs, including acoustical panels and chambers to block out ambient noise. When a worker moves into a chamber, the annoying sound of telephones becomes practically inaudible. Ambient noise was further controlled by sound-absorbing ceilings and speakers that introduced background white noise on each floor.

The workstations were also designed to be flexible and fast-cycle, allowing varying configurations as well as rapid reconfigurations. The floor is raised ten inches, which provides for unlimited underfloor wire-way systems. According to Russell Robinson, manager of facility interior development, 70 percent of Lockheed's employees are moved every year. In older buildings reconfiguring each work area costs up to $600 per change. Moves in Building 157 involve "unplugging equipment, placing two piano dollies under the work station and relocating it," according to *Facilities Planning News*. "The whole process takes approximately one-half hour and costs $60 per work station."

Employees love the building. More than a year after occupancy, a survey of workers included the following responses:

EMPLOYEE REACTION TO BUILDING 157

I love my work station. It's very comfortable, quiet, pleasant, with good lighting, air conditioning—it's great. . . . I think the daylighting is a major contributor to the pleasantness of the surroundings—it's a very comfortable place to work.

—Jacques Avedissian, design engineer

My work space is 15 feet from the litetrium and the lighting is great. The office decor, arrangement, and temperature are ideal. There are many people working on this floor, but the feeling is not one of crowding, but of spaciousness. Interface with other departments is greatly facilitated because we're finally all in one building. By nature I'm very cynical, but the conditions in this building are far superior to any I've experienced in 30 years in the aerospace industry.

—Ben Kimura, staff engineer

I love my work space. I think the building itself is very pretty; my own work station is very functional. I am five work stations from the window and the light is fine. I use my task light and could order an additional desk lamp if I felt the need to. I like the daylight.

—Joanne Navarini, financial controller

Robinson reports that "productivity is up" because absenteeism was reduced. Lockheed itself never published the figures concerning the improvements in absenteeism and productivity, but, according to Don Aitken, then chairman of the Department of Environmental Studies at San Jose State University, "Lockheed moved a known population of workers into the building and absenteeism dropped 15 percent." Aitken led numerous tours of Building 157 after it opened and was told by Lockheed officials that the reduced absenteeism "paid 100 percent of the extra cost of the building in the first year."

Lee Windheim also reports that Lockheed officials told him that productivity rose 15 percent for the first major contract done in the building compared with previous contracts done by those Lockheed engineers. Aitken reports an even more astonishing anecdote: Top Lockheed officials told him that they believe they won a very competitive $1.5 billion defense contract on the basis of their improved productivity—and that the profits from that contract paid for the entire building. Lockheed Building 157 may be the most lean and clean building ever built.

WAL-MART

In June 1993, a new prototype Wal-Mart store opened in Lawrence, Kansas. Called the Eco-Mart, the building is an experimental foray into sustainable design by the nation's largest retailer. The project was led by Wal-Mart's Environment Committee and BSW Architects of Tulsa, Oklahoma. The design consulting team involved a number of firms, including the Center for Resource Management, William McDonough Architects, and the Rocky Mountain Institute. The team introduced a series of environmentally responsive design strategies and technologies.[8]

Elements of this experiment include the use of native species for landscaping; a constructed wetlands for site runoff and as a source for irrigation; a building shell design for adaptive reuse as a multifamily housing complex (for when the building is no longer used as a store); a structural roof system constructed from sustainably harvested timber; an environmental education center; and a recycling center. A major goal of the project was to design for energy efficiency. The building has a glazed arch at the entrance for daylighting, an efficient lighting system, a heating and cooling system that utilizes ice storage, and a new type of light monitor (advanced skylights) developed specifically for this project.

The building had some kinks to work out. The controls on the lighting systems were not compatible with the ballasts. The ice storage system leaked water, and, due to the expanded hours of store operation, was not able to fully refreeze during the off-hours in the evening to provide complete cooling during the day. The energy performance of the building is, however, better than other Wal-Marts, and will be better still when these problems are solved.

Wal-Mart's normal costs are extremely low, and a building typically pays for its own construction costs in a few months. The Eco-Mart costs about 20 percent more than Wal-Mart's normal construction cost per square foot, for several reasons: the sustainably harvested timber added 10 percent to the roof cost, the integration of systems was not optimized resulting in a more expensive cooling system, and the building contains elements not found in other stores. One of these is the light monitors. As a cost-saving measure, Wal-Mart decided to cut the number of light monitors in half. Rather than scatter the monitors across the roof, they were placed on only half the roof, leaving the other half without daylighting.

These conditions created an unintentional experiment whose results have caught the attention of the corporation's management. Each of Wal-Mart's cash registers is connected in real time back to the headquarters in Bentonville, Arkansas, as part of the company's highly effective "just-in-time" retail stocking and distribution system, which allows them to respond rapidly to changing customer demands. According to Tom Seay, Wal-Mart's vice president for real estate, this allowed the company to discover "that the sales pressure [sales per square foot] was significantly higher for those departments located in the daylit half of the store." This sales rate was also higher than the same departments in other stores. Additionally, employees in the section without the light monitors are arguing that their departments should be moved to the daylit side. Wal-Mart is now considering implementing many of the Eco-Mart measures in both new construction and existing stores.

WEST BEND MUTUAL INSURANCE

West Bend Mutual Insurance Company's new 150,000-square-foot headquarters in West Bend, Wisconsin, was the subject of one of the most carefully documented increases in productivity from lean and clean design. The building won the 1992 Intellex Building for Excellence Award, cosponsored by *Consulting-Specifying Engineer* magazine and the Intelligent Buildings Institute. The building makes use of a number of energy-saving design features. Energy efficiency was incorporated into the lighting system (which includes task lighting and occupancy sensors), windows, shell insulation, and HVAC system, which uses a thermal-storage system that makes ice electrically overnight to help cool the building during the day. These measures allowed West Bend to get utility rebates that kept the project within its ninety-dollars-per-square-foot budget.[9]

In the West Bend Mutual building, enclosed offices all have individual temperature control. But the most high-tech features of the building are the Environmentally Responsive Workstations (ERWs). Workers in open-office areas are given direct, individual control over both temperature and airflow. Radiant heaters and vents are built directly into their furniture and controlled by a panel on their desks, which also provides direct control of task lighting and of white noise levels (to mask out nearby noises). A motion sensor in each ERW turns it off if the worker leaves the space and brings it back on when he or she returns.

By giving workers direct control over their environment, the ERWs allow individuals working near one another to have very different temperatures in their spaces. No longer need the entire HVAC system be driven by a manager, or by a few vocal employees who want it hotter or colder than everyone else. The motion sensors save even more energy. It is worth noting that before the move into the new building, West Bend Mutual employees were given the chance to try out and comment on a full-scale mock-up of the ERWs. Those workers who had expressed the greatest doubts about the ERWs were allowed to test them at their own desks.

The lighting in the old building had been provided by overhead fluorescent lamps. The workers in the new building all had task lights, and they could adjust them according to their preference for brightness. The annual electricity costs in the old building were $2.16 per square foot. The annual electricity costs in the new building are $1.32 per square foot. The $0.84 per square foot savings—a 40 percent reduction—is particularly impressive given that the old building got its heat from gas-fired boilers whereas the new building is completely electric.

The Center for Architectural Research and the Center for Services Research and Education at Rensselaer Polytechnic Institute (RPI) in Troy, New York, conducted a detailed study of productivity in the old building in the twenty-six weeks before the move and in the new building for twenty-four weeks after the move. The RPI study employed a productivity assessment system used by West

Bend Mutual for many years, which basically tracked the number of insurance files processed by each employee per week. Researchers also conducted a detailed tenant questionnaire survey of workers' perceived levels of comfort, air quality, noise control, privacy, and lighting before and after the move. The study concluded, "The combined effect of the new building and ERWs produced a statistically significant median increase in productivity of approximately 16% over productivity in the old building."[10]

In an attempt to determine just how much of the productivity gain was the result of the ERWs, the units were turned off randomly during a two-week period for a fraction of the workers. The researchers concluded, "Our best estimate is that ERWs were responsible for an increase in productivity of about 2.8% relative to productivity levels in the old building." This figure almost certainly underestimates the actual benefit of the ERWs, according to Ronald W. Lauret, West Bend Mutual's senior vice president. Lauret observes that many workers demanded that their units be turned back on immediately. Some even threatened to go home (they were eliminated from the study). He estimates that if those employees were factored back in, the productivity gain from the ERWs alone would have been 4 percent to 6 percent.

The company's annual salary base is about $13 million, so even a 2.8 percent gain in productivity is worth about $364,000. With Lauret's higher estimate, the ERWs paid for themselves in under one year, and, in his words, "That's a substantial return on investment."

DESIGN FOR PRODUCTIVITY

Several general conclusions can be drawn from the case studies here and in earlier chapters. First, energy-efficient design can have a significant impact on productivity. Improvements of 7 to 15 percent can be achieved. At the same time sick leave and absenteeism may drop significantly.

Second, productivity gains are typically accompanied by fewer errors and defects. Many errors are the results of visual problems that state-of-the-art lighting can resolve. Simply put, *quality lighting and quality work are inseparable.* It might also be said that pollution prevention and defect prevention go hand in hand.

Third, people like daylight. This was as true for Lockheed's employees as for Compaq's. The more connection to the natural environment, the better. Sociobiologist Edward O. Wilson coined the term *biophilia* to describe what he believed is our innate affinity for the natural world. A 1993 book, *The Biophilia Hypothesis*, examines the evidence in detail and cites studies of hospitals and prisons that "suggest that prolonged exposure to window views of nature can have important health-related influences." Roger Ulrich, associate dean for research at Texas A&M University's College of Architecture and an expert in behavioral geography and environmental psychology, wrote in the book:

> Research on biophilia is at a relatively early stage of development, and no findings have yet appeared that constitute convincing support for the proposition that positive responding to nature has a partly genetic basis. Perhaps the most persuasive findings currently available are the striking patterns across diverse groups and cultures revealing a preference for everyday natural scenes over urban scenes lacking nature.[11]

Biophilia or not, daylight is the most cost-effective and least polluting way to light, and heat, a building.

Fourth, the two elements of lean and clean design that most improve productivity and quality are (1) focusing on the end user and (2) giving workers more control over their environment. Focusing on the end user—a basic systems principle—here means designing lighting with the worker's task in mind. This invariably saves energy, and its impact on worker performance can be substantial. Giving workers more control over their workplace environment—adjustment of lighting, heating, and cooling—makes them more comfortable and more effective, and it reinforces the best Japanese (and American) management practices of Total Quality Management and fast-cycle manufacturing.

Fifth, a systems approach to *new* building design achieves the most cost-effective improvements in energy use and productivity. Designing a building right the first time is obviously cheaper than retrofitting an inadequate design. For one thing, you don't have to buy equipment twice. The best lean and clean gains are achieved when green design principles are accompanied by other techniques aimed at boosting productivity, such as improved acoustics. In the NMB Bank, Lockheed Building 157, Wal-Mart's Eco-Mart and West Bend Mutual, a variety of measures were taken. We can never absolutely measure the contributions of green design principles— indeed, the point of a systems approach is that the whole is greater than the sum of its parts. But employee and management comments, as well as a detailed independent study (West Bend Mutual), make clear that green design is a major contributing factor to the productivity jump.

Sixth, while designing an office right the first time is best, upgrades can be very effective. This conclusion is particularly critical because the vast majority of businesses will not be moving anytime soon, and, even when they do, most will not be commissioning a whole new building. The cases of the Reno Post Office, Boeing, and the two drafting areas make clear that even simple improvements in lighting can create tremendous savings.

Seventh, and finally, savings from the gain in productivity will dwarf savings in energy, in some cases by a factor of ten. Payback in a few years becomes payback in less than one year. Return on investment can jump from 25 to 500 percent.

7

Becoming Lean and Clean: A Systems Approach

In the 1990s, you can go from market dominance to decay in a couple of years. Only the nimble can avoid that fate. Being nimble means avoiding multiple layers of management, keeping close to your customers and putting a premium on quality and service. Nimble firms anticipate new markets and eliminate time-wasting steps from product development, manufacturing and marketing.

—*Forbes*, January 4, 1993

Becoming lean and clean requires a fundamental reorientation of a company's worldview or paradigm. When most of a company's workers and managers are unable to adapt, all change is thwarted. Indeed, if the shift in behavior had to be as radical as the shift in thinking, change might well be impossible. No company can change its behavior—its processes and products—overnight. Few people can either. Fortunately, while the necessary change in corporate culture is revolutionary, the change in product, process, and employee behavior is evolutionary; it can be accomplished with constant, incremental improvement, which is the essence of kaizen, Total Quality Management, lean production, and fast-cycle management. In these pages fast-cycle management is not merely a way to think about how organizations change; it is the change itself. This chapter and the next synthesize the improvements a company, particularly a manufacturer, must make to become fast-cycle with those needed to prevent pollution.

Consider Mitsubishi's heat pump, which the authors of *Compet-*

ing Against Time offer as a prime example of fast-cycle manufacturing.[1] Every year between 1979 and 1988 the Mitsubishi Electric Company added a new feature or made a major design change in its three-horsepower heat pump, including the introduction of integrated circuits to control the pump cycle (1980), microprocessors (1981), and an inverter for controlling motor speed (1984).

Not until the mid-1980s did an American company even consider the use of integrated circuits in its residential heat pump. It would have taken four to five years to bring the product to market, and even then the U.S. firm would only have had in 1990 a product comparable to the 1980 Mitsubishi heat pump. The American company threw in the towel, purchasing its advanced air conditioners, heat pumps, and components from the Japanese competition.

Not only did the fast-cycle strategy win, but most of the design changes improved the energy efficiency of the heat pump, reducing its environmental impact. And the microprocessors, together with "quick connect" Freon lines and simplification of the wiring, made the product simple to install and very reliable. This is lean and clean design.

Computers and telecommunications have collapsed time frames for everyone, but most of all for those in business. Speed is the watchword of our time. If there is an element common to these faltering corporations and their leaders it is that they could not or would not adjust to the new reality of collapsed time frames.

—*Wall Street Journal* editorial on the troubles at GM, IBM, American Express, and Sears[2]

Tachi Kiuchi, chairman and CEO of Mitsubishi Electric America, says that the company has made a "conscious effort" to combine the goal of achieving competitive advantage with the goal of improving the energy and resource efficiency of its products (and its manufacturing processes and operations). Kiuchi notes two advantages Japanese companies have in reconciling environmental and competitiveness goals: First is their expertise in process improvement, which is essential to clean production. Second is Japan's scarcity of resources, which has led companies to seek added value through knowledge-based skills such as efficient design. For Kiuchi, lean production and clean production are wholly compatible.[3]

Companies that don't become fast-cycle will be passed by. Companies that become lean and clean by integrating fast-cycle and pollution prevention will leapfrog the rest.

THE IMPORTANCE OF FAST-CYCLE

Colonel John Boyd was a renowned air force pilot in the Korean War. He was puzzled by the fact that in that war the American F-86 Sabre jet consistently beat the Soviet-built MiG-15 in aerial dogfights, even though the MiG was a "superior" plane by traditional standards: It could accelerate more rapidly, climb faster, and generally turn tighter than the F-86.[4]

Boyd discovered that the F-86 had two crucial advantages. First, the jet's glass-domed bubble canopy enabled the pilot to observe enemy activity more easily than did the MiG cockpit. Second, the F-86 could change from one maneuver to another much more quickly—decelerating and diving while turning or switching directions. Thus, the F-86s' better *observation* and faster *maneuvering* canceled the superiority of the MiGs. As the planes danced around each other in the classic dogfight, the F-86 would gain a larger and larger positional advantage, rapidly changing from potential victim to predator.

All successful competitive organizations—in business or warfare—have one advantage in common: They act more quickly than their competitors. The point may seem obvious, yet even as recently as the late 1980s, most U.S. businesses had not accepted the importance of focusing on time. The bias toward operations over process, and the lack of dynamic systems thinking, slowed U.S. companies from seeing the importance of time. Today the mainstream business press and the academic literature have finally embraced time. *Forbes* endorses "nimble" companies; *The Wall Street Journal* proclaims, "Speed is the watchword of our time." Yet few are clear about what should be speeded up.

A 1990 study in the *Sloan Management Review* that reviewed

data from more than 400 U.S. and foreign companies concluded that cutting product cycle time in half adds two or three percentage points to a plant's rate of productivity growth. Out of many possible ways to improve productivity, the only ones that "were statistically shown to be consistently effective" were related to just-in-time:

> The plants with the highest productivity gains were not, in general, distinguished from those with lesser gains by heavy investment in high tech, . . . gain-sharing plans, or conventional industrial engineering approaches that concentrate on the operator's job. The only consistent distinctions were those relating to JIT [just-in-time], specifically, [cycle] time reduction, improved quality, lower inventories, and participative techniques.[5]

In other words, whereas time is crucial for improving productivity gains, speeding up the operator's job is not. The distinction is between process and operations.

Reducing cycle time is not another expensive management fad; it is a necessity. If you don't do it, you can be sure your competition will. One 1993 study reported cycle time reductions in twenty-one new product introductions.[6] Here are just a few:

PRODUCT	CYCLE TIME (MONTHS)	
	Previous	*Current*
Construction equipment	84	50
Medical imaging equipment	72	36
Cars	60	36
Copiers	60	36
Trucks	60	30
Personal computers	48	14
Printers	54	22
Thermostats	48	10
Air-powered grinders	40	15
Electronic pagers	36	18
Electric clutch brakes	39	9
Lanterns	24	12
Cordless phones	24	12

Manufacturers are not the only ones cutting cycle time. Banks, brokerage firms, insurance companies, utilities, cable companies, retail stores, and hotels—the entire service sector—are learning how to eliminate wasted steps in their work. The issue is no longer *why* a business should reduce cycle time but *how*—especially since any systematic approach to reducing wasted time will also help an organization reduce wasted resources.

THE O-O-D-A LOOP: OBSERVE, ORIENT, DECIDE, ACT

The systematic time-based approach developed by Col. John Boyd can and should be integrated with a systematic approach to pollution prevention. Used by the Allies in the Persian Gulf War, Boyd's approach was a key reason that we won far more quickly, and with far fewer casualties, than virtually everyone expected. The power of that approach was made clear in a midwar briefing by marine Brig. Gen. Richard Neal, deputy director of operations for U.S. Central Command. Describing U.S. strategy against Saddam Hussein, Neal said, "We're inside his decision-making cycle. . . . We're kind of out-thinking him. . . . We can see what he's been doing, we can kind of anticipate what his next move is going to do, and we can adapt our tactics accordingly."[7]

After the Korean War, Boyd studied engineering and military history to learn how to generalize his experience into a coherent theory of warfare: "A Discourse on Winning and Losing," as he ultimately called it. He derived much of his thinking on fast-paced maneuver warfare from Sun Tzu, the brilliant Chinese military strategist of the fourth century B.C. In his classic book, *The Art of War*, Sun Tzu wrote, "To win one hundred victories in one hundred battles is not the acme of skill. To subdue the enemy without fighting is the acme of skill."[8] The book has been widely used for more than 2000 years by military commanders and is a favorite among Japanese businesspeople. It has only lately been picked up by Americans.

The key to winning, Boyd concluded, is to operate at a faster tempo than the adversary, to get inside what he called their O-O-D-A loop. The loop consists of *observing* the competition's actions, *orienting* oneself to the unfolding situation, *deciding* what to do, and then *acting*. The action (and any response it evokes) then alters the situation, necessitating new observations and a repetition of the cycle. When one side's O-O-D-A loop is faster, shorter, or more efficient than its opponent's, the first will run circles around the second. The slower side will be constantly reacting to its adversary's previous moves, unable to take the initiative.

This is exactly how Japanese companies used flexible manufacturing to compete so well in the 1980s: They got inside other companies' time-cycle loops. In industries such as automobiles, air conditioners, and projection televisions, they can cycle through the entire production system—research, development, marketing, production, and sales—in one-half to one-third the time of companies from any other nation, usually with a comparable reduction in cost.

At double or triple speed, Japanese companies achieve an enormous advantage with quick innovation through a process of steady, incremental improvements that keep them at least a step ahead of their competitors. And when the improvements also focus on environmental impact, the result can be a lean and clean design, as in the case of the Mitsubishi heat pump. Here are specific steps that can minimize the inefficiency and waste in each of the four phases of the Observe-Orient-Decide-Act loop and create a lean and clean business.

I. IMPROVING OBSERVATIONS

Whoever best observes the changing world can best respond to it. Rigorously observing the world is the only way to avoid being trapped in an outdated worldview. "We learn to rely on our concepts of reality more than on our observation," wrote Robert Fritz in *The Path of Least Resistance*. "It is more convenient to assume that reality is similar to our preconceived ideas than to freshly observe what we have before our eyes."[9]

To get the best, freshest observations, one must be as directly connected to one's business environment as possible and get constant, rapid feedback from employees, customers, and the competition.

1. Talk to Customers
This is a credo of Total Quality Management. Talking to customers is the most important observation—the best feedback. Quality is in the eye of the beholder, the customer. Focus on the end result: happy customers. Yet only 22 percent of top U.S. companies always or almost always use customers' expectations to develop new prod-

ucts and services, compared with 40 percent for German companies and 58 percent for Japanese.[10]

Customers' perceptions and misperceptions determine how and why they use—and misuse—a product. Talking to them is crucial for figuring out how to improve a product and for ensuring that they are as satisfied as possible. Talking to them is crucial for shattering myths. Consider the air conditioner. Most people, including many with technical training, believe that the thermostat on an air conditioner works like a valve: the higher or lower you set it, the *faster* the room will heat up or cool down. It will not. But many people walk into a warm room and turn the thermostat to 50 degrees. Then, of course, when it gets too cold, they turn it way up. Maybe you do this, even though it can waste energy, result in *less* comfort, and wear out the equipment faster.

If so many people misuse one of the most common products in the world, imagine how they misuse your product. If so few people bother to program the time on their VCRs, even though it really isn't that hard if one reads the manual, what do they do with your product and manual? When your designers, engineers, and workers talk to customers, and observe them using your product or a competitor's, they can design the product your customers want, one that is easy to use correctly.

The ideal: Every person in a company should visit customers every year. One hundred ideas detailed for improving quality internally are nothing compared with regular communication with customers. As noted in Chapter 2, listening to customers has lead Xerox to emphasize an integrated approach to extended life, reduced cost, recycling, and remanufacturing.

2. Benchmark Against the Best

Observe what your toughest competitors are doing and benchmark yourself against them. You will have to stretch your current thinking, since your competitors may approach problems differently than you do. Indeed, things you may consider impossible (such as reducing cycle time by 90 percent, scrap by 90 percent, or energy use by 50 percent) might be their standard operating practice. Don't view their unusual practices as eccentricities; instead, consider whether they may be essential to a new and better approach. Talk to some of their customers. What do they want you to do to win them over? Motorola has an "intelligence department" that monitors conferences, journals, and industry gossip to track the latest technology developments, anticipate breakthroughs, and stay ahead of the competition.[11]

Environmental benchmarking is also important. Don't find out merely what the biggest companies are doing in terms of cycle time, defects, or technology development. The biggest companies are not always the best. Find out who has reduced waste the most, who has the best recycling program, who uses the least energy. This book offers some idea of what the best have done in energy efficiency and pollution prevention, but statistics in books are no substitute for visiting and observing an efficient plant. Benchmarking trips allow managers and employees to observe directly what the competition has done and see for themselves that business will suffer if they don't match that performance.[12]

3. Measure

Measuring is rigorous observation. The mantra of Total Quality Management is "you only get what you measure." If you don't measure customer satisfaction, how will you be able to increase it? If you don't measure cycle time, defects, scrap, energy use, resource use, pollution, and the like, benchmarking will be pointless. You won't have a baseline of comparison. You won't be able to know if changes you make have any effect. You won't be able to tell your people how much they have improved. And you won't be able to tell outsiders either.

Choosing what to measure is not simple. There are a thousand potential quantities to track. A company's vision, orientation, or worldview—to be discussed shortly—will most affect what it measures.

4. Talk to Employees

Managers must know what line workers know, since employees know where all the waste is. Work with them to improve the processes that will cut that waste in half within three years, then in half again in another three years.

Find out how workers regard the environment. Find out what they are doing at home and in the community to improve the environment. Trying to impose environmental consciousness—energy efficiency and clean production—on workers will lead to resistance, and environmental consciousness will go the way of most management fads. Recognize that many workers may well be further along in their own lives than your company is in its operations. In particular, the youngest and newest employees are probably the most interested in the environment and the most critical of your company's policies.

II. COMMON ORIENTATION

Orientation, the second *O* in the O-O-D-A loop, is the most crucial element of the loop. No matter how efficiently an organization observes, decides, and acts—no matter how fast its cycle time—it is doomed to fail when it is headed in the wrong direction, whether building cars no one wants or, if an army, attacking in the wrong place. The rise of most great organizations can be traced to their embracing a powerful, unifying purpose. And their fall can be traced to their having lost that clear sense of purpose after they achieved success or failing to change to a new unifying purpose as the world changed.

When John F. Kennedy set his ambitious goal to put a man on the moon by the end of the 1960s, he gave NASA a powerful vision. Managers at NASA further embraced the idea of prevention—accident prevention. Every employee, from top management to line engineers, gave priority to safety. This orientation helped make NASA one of the most resilient, problem-solving, and successful federal agencies.

By the mid-1980s, however, NASA had lost much of its focus. To economically justify its primary product—the space shuttle—as a launch vehicle, it had oversold the shuttle's capability. The agency was under tremendous pressure to keep up a demanding launch schedule. In January 1986 management overrode warnings from engineers that it was too cold to launch. The O-ring cracked. The result: the deaths of seven truly remarkable Americans, a $2 billion loss in equipment, and an incalculable shock to the nation.

Orientation is the lens through which we view the world. It is our paradigm, vision, and philosophy. It is our mental model—our "internal pictures of how the world works," as Peter Senge puts it.[13] When one's internal pictures do not match the external world, only failure can follow. Consider the beliefs GM developed in the postwar years, according to management consultant Ian Mitroff:[14]

- Cars are primarily status symbols. Styling is therefore more important than quality.

- The American car market is isolated from the rest of the world.

- Workers do not have an important impact on productivity or product quality.

- Everyone connected with the system has no need for more than a fragmented, compartmentalized understanding of the business.

Similar myths in the guise of principles have governed much of American business in the postwar era. These beliefs worked quite well when the war-ravaged competition was still recovering from World War II. But the auto industry considered its beliefs "a magic formula for success for all time, when all it had found was a particular set of conditions . . . that were good for a limited time."[15]

After nearly two decades of inaction, many U.S. companies are reorienting. They have begun a paradigm shift concerning quality, international competition, worker input, and cross-functional teams—but they have been slow to adopt energy efficiency and clean production. I have found a set of misguided beliefs about the environment that are held as firmly today as the preceding GM list was in the past:

BUSINESS-ENVIRONMENT MYTHS

- Increasing production and profits requires increasing energy and resource use.

- Pollution is a natural result of business operations.

- Efforts to reduce either resource use or pollution can only hurt the bottom line.

- Environmentalism is a fad best dealt with through better public relations.

- The workplace environment has no important impact on worker productivity.

We cannot wait another two decades for U.S. companies to realize that these beliefs are false. A key goal of this book is to shatter these myths. But merely holding up some case studies of "green" businesses offers little hope of creating enduring change in the absence of a compelling new explanation within a conceptual framework. And a conceptual framework is useless without a specific set of practical steps for beginning the permanent process of incremental change.

Why is it so difficult for a CEO, a company, or an entire industry to reorient, to change mental models? One central problem is that any flawed orientation or mental model is self-validating and self-perpetuating. Inconsistencies between one's mental model and reality are uncomfortable, so human nature rescues us from the discomfort of this "cognitive dissonance." Rather than change the mental model, we discount the information that would create unease.

In the late 1980s the Massachusetts Institute of Technology conducted the most comprehensive survey study ever undertaken of the global auto industry. This $5 million, five-year, fourteen-country study culminated in a seminal book, *The Machine That Changed the World.* The survey began in 1986 at GM's Framingham, Massachusetts, assembly plant, a traditional U.S. manufacturer. The first interview is a classic case of cognitive dissonance, of the inability to see anomalies:

> The plant's senior managers . . . had just returned from a tour of the Toyota-GM joint-venture plant (NUMMI). . . . One reported that secret repair areas and secret inventories had to exist behind the NUMMI plant, because he hadn't seen enough of either for a "real" plant. Another manager wondered what all the fuss was about. "They build cars just like we do." A third warned that "all that NUMMI talk [about lean production] is not welcome around here."[16]

A rigid orientation can undermine even practices as essential as benchmarking.

A company that wants to respond to unfolding events will have to force itself constantly to face unpleasant facts. But it is never enough to seek out only information that might undermine current thinking, no matter how committed the search is. A successful company will have to institutionalize these mechanisms:

- Examining its mental models
- Comparing them to reality

- Considering whether a new paradigm might be needed

- Adopting it, when necessary

The Royal Dutch/Shell oil company has probably done this more successfully than any other company in the world.

ROYAL DUTCH/SHELL: REORIENTING *BEFORE* THE OIL CRISIS

In the early 1970s Royal Dutch/Shell's planning staff anticipated the oil price shocks that occurred in 1973 and 1974. Shell moved from being the weakest of the seven largest oil companies in 1970 to one of the two strongest ten years later.[17]

Correctly anticipating the future was not the hard part for the Planning Group. One codeveloper of Shell's planning process, Pierre Wack, has written, "Surprises in the business environment almost never emerge without warning." Many others foresaw the oil crisis. The hard part was getting Shell's managers to rethink their mental models. Wack came to realize that providing new information was not enough, because "novel information, outside the span of managerial expectations, may not penetrate the core of decision makers' minds, where possible futures are rehearsed and judgment exercised."[18]

Wack compares that time to the days before the attack on Pearl Harbor, when there was a massive volume of intelligence signals ("noise") coming in. He quotes Roberta Wohlstetter writing in 1962: "To discriminate significant sounds against this background of noise, one has to be listening for something or for one of several things. . . . One needs not only an ear but a variety of hypotheses that guide observation." The Japanese commander of the attack, Mitsuo Fuchida, was quite surprised he had achieved surprise. Before the Russo-Japanese War of 1904, the Japanese Navy had used a surprise attack to destroy the Russian Pacific fleet at anchor in Port Arthur. Fuchida asked, "Had these Americans never heard of Port Arthur?"[19]

The approach Wack developed at Shell was "scenario planning," but it was a type of scenario planning entirely different from that of most companies. Wack did not want merely to quantify uncertainties—i.e., the price of oil may be twenty or forty dollars per barrel in 1995—because doing so offers little help to decision makers. Wack wanted to offer managers two or more complete worldviews or scenarios—grounded in a sound analysis of reality. One of these scenarios might be business as usual. At least one would be a radically different, though plausible, view of the world.

Even though Wack foresaw the energy crisis and presented the results to Shell's management, "no more than a third of Shell's critical decision centers" were acting on the insights gained from the energy crisis scenario. Wack came to realize that although all managers had the new information, most were still processing it through their old paradigm or mental model, what Wack called their microcosm. He says,

> I cannot overemphasize this point: unless corporate microcosm changes, managerial behavior will not change; the internal compass must be recalibrated. . . .
>
> Our real target was the microcosms of our decision makers: unless we influenced the mental image, the picture of reality held by critical decision makers, our scenarios would be like water on stone.

Wack and his fellow planners realized they "needed to design scenarios so that managers would question their own model of reality and change it when necessary, so as to come up with strategic insights beyond their minds' previous reach." The Planning Group designed a set of scenarios early in 1973 that would force a paradigm shift. Shell managers were presented a business-as-usual scenario. They were also presented the underlying assumptions required for that scenario to hold. Those assumptions were shown to be wholly unrealistic, requiring several "miracles," each of which was highly improbable, to occur simultaneously. The only way to delay the energy crisis would be the discovery of "new Middle East–sized oil reserves in an area that would have no problem in absorbing revenues" or "seizure and control of producers by consuming countries."

Once managers saw that their faith in the status quo was built on miracles, they were more receptive to new thinking. Wack and his fellow planners led the Shell managers through the process of building a new paradigm, showing them what was likely to happen in the future and what its implications were for the managers' own decisions and actions.

Since oil price increases were inevitable, oil demand would drop. Demand would no longer outpace GNP. Using this scenario, Wack told the refining managers to prepare to become "a low-growth industry." Not an easy message to hear.

The planners made clear that the energy shock would have dramatic effects worldwide but would affect individual nations differently. The effects would depend on whether a given country was an oil exporter or importer, whether it was free market or centrally planned, and, for importers, how much they relied on imports and how easily they could find alternatives. Therefore, one basic, rigid strategy would not be useful for operating different companies in different parts of the world. Each region would have to respond independently. As a result, Shell would "need to further decentralize the decision-making and strategic process."

Even those managers who remained skeptical at least understood the flaws in their old paradigm and the powerful implications of the new one. When the OPEC oil embargo did occur, and the underlying assumptions of the energy crisis scenario were proven correct, Shell managers were far quicker to shift their behavior accordingly. They slowed down investments in refineries. Their projections of energy demand were consistently lower and more accurate than their competitors'. They decentralized, while their competitors were becoming more centralized—and hence more inflexible—in a world of rapidly changing events.

Shell rose rapidly from its position as the weakest of the giant oil companies. Chapter 9 will examine the lean and clean possibility that Shell's strategic planners now see as one future for the world. A company will not change in a significant or enduring fashion if it does not systematically change its managers' worldview, if it does not change its orientation. But such change is not easy. Here are some measures that may help.

1. Relentlessly Seek Different Views Externally
Everyone in the company should read *The Art of the Long View* by Peter Schwartz (former Shell planner, now head of the Global Busi-

ness Network); *The Ecology of Commerce: A Declaration of Sustainability* by Paul Hawken; *The Fifth Discipline* by Peter Senge; and *Beyond the Limits* by Donella Meadows et al. General Alfred Gray had heard Colonel Boyd's briefings; after taking command of the marines in the late 1980s, Gray made Sun Tzu's *The Art of War* required reading for all marines.[20]

Bring in an outside expert or team to spend a day going through Shell's scenarios with your top management. I suggest Peter Schwartz or Amory Lovins (director of research at Rocky Mountain Institute) or Peter Senge.

Talk to the Global Environmental Management Initiative, based in Washington, DC. Formed in 1990, it is a partnership of twenty-three major companies, including AT&T, Boeing, Dow, Kodak, Du Pont, GE, Merck, Procter & Gamble, Tenneco, and Union Carbide. It is committed to spreading Total Quality Environmental Management. Its fundamental belief is simple:

Business, by taking control of its environmental destiny, can spur change from within and create a forum for sharing solutions among industry worldwide.

—Global Environmental Management Initiative

---

2. Encourage Dissent Internally

According to *The Wall Street Journal,* "Motorola makes a cult of dissent and open verbal combat. Employees are entitled to file a 'minority report' if they feel their ideas aren't being supported. The reports are read by bosses of the teams' bosses." Retribution for such dissent is "considered 'unmacho' or craven."[21]

Karney Yakmalian, a materials manager at Motorola, puts it differently: "The environment and culture is such that you are encouraged to strive to find better ways of doing things." That, in turn, creates the need for an "environment to constructively comment on the way things are done." There is "no feeling you have to concern yourself with the impact of criticism." Instead, there's a "feeling of ongoing respect for colleagues."[22]

Improving constantly requires a culture that fosters dissent.

Long-term success demands that the dissent be constructive, not destructive. Still, the medicine can be bitter. In 1979, at a gathering of top Motorola officers and their spouses, one product manager stood up and declared, "I think our quality stinks. Some of our competitors are much better and we ought to do something about it."[23]

Motorola did embrace quality. Today it builds some of the highest quality products in America. The company has reduced defects 90 percent or more in some products and slashed product cycle time in half. In 1987 it set a goal to reduce the defect rate to 3.4 per million parts (known as six-sigma quality) by 1993. From 1987 to 1992 Motorola nearly doubled productivity and earnings but achieved "only" 30.0 defects per million. It not only expects to meet the 3.4 defects per million goal but is already thinking about a defect rate measured in parts per billion.

3. Encourage Staff to Challenge Basic Assumptions

Ken Teeters, plant engineering manager at Harrah's Hotel and Casino in Las Vegas, had, over many years, been improving Harrah's energy efficiency. The tortoise approach was necessary because, as Teeters relates, a casino measures every investment against a new slot machine, which pays for itself in three months. A solid case had to be developed for each efficiency investment, and then the savings had to be carefully documented. As Teeters puts it, "Always quantify the expected savings and show how the savings will be achieved. Then track the energy savings results and provide feedback to senior management."[24]

Teeters's energy management team realized that in an 1800-room hotel such as Harrah's, changing the sheets every day uses a tremendous amount of water, energy, and detergent. But how do you change one of the most time-honored hotel traditions—fresh linen every day?

Teeters's team first asked themselves if they changed their sheets at home every day. The answer to this minibenchmarking exercise was uniformly no. But the assumption was still that people expected better service at a hotel than they did at home. "They want to be pampered. They want a 'nonhome' type experience." That's why they pay high room rates. Teeters's team was able to overcome that obstacle with a simple pilot program:

We have a whole environmental story and energy savings story to tell. So we put it on a flyer and we put it nicely on the pillow each day. And basically we said, "If you're a stayover guest and if you want your linens changed daily—not your towels and everything—you have to be *proactive* to call and request it."

The response: "The guests love it." About 95 percent said they were glad to have the choice, and the vast majority chose not to have their sheets changed every day. The result isn't that surprising: The average stay at Harrah's is only 3.4 days, and hotel guests are concerned about the environment, too.

Harrah's saved $70,000 in energy and water costs the first year. The environmental benefits include reduced use of energy and water, reduced air pollution, and far less release of dirty, soapy water. Harrah's reduced the amount of linen processed each month by 100,000 pounds. Sheets last longer now, too. The end-use approach—asking people what they want and then giving it to them—invariably costs the least.

The idea may seem obvious in terms of profits, customer service, and the environment. Indeed, it may seem ridiculous for a hotel in the middle of the desert to wash 1800 sheets a day and not even ask its guests if that's what they want. Yet it is standard operating procedure for virtually every hotel in Nevada, or the United States for that matter. It's tradition.

Such win-win opportunities are virtually unlimited. For a company to find them, employees must be encouraged to challenge the most basic assumptions of the business. In Teeters's words, "You've got to be an innovator. In today's world, if you are afraid of technology and change, you put yourself at risk."

4. Embrace Life-cycle Analysis

One of the best ways to reorient is to calculate *all* the costs and benefits of your actions. This life-cycle analysis, discussed earlier, inevitably leads to energy efficiency and pollution prevention. It also leads to customer service. Home Depot, a giant discount outlet, calculates that typical shoppers spend $38 a visit but visit the store thirty times a year, ultimately spending $25,000 each throughout their lifetime. If Home Depot treated each shopper as worth a mere

$38, service might be terrible. The life-cycle analysis, however, drives Home Depot to treat each shopper as worth $25,000. Such a customer must be pampered. Others studies also indicate that spending on customer service reduces life-cycle costs.[25]

5. Experiment with Green Office Design on One Floor

Take a benchmarking trip to one of the green offices discussed in Chapter 6. Talk to the line workers and the managers. Talk to the architects.

Ask the architects to design a renovation for one of your floors. Pick a floor where you can accurately measure productivity. Have some of the employees on the floor work with the architects. Track productivity for a year. Observe for yourself that productivity jumps and then stays at the higher level.

6. Train Everyone in the New Orientation

Once top management decides upon the lean and clean vision, all employees must be trained in it. Such training in a common orientation was crucial in the Persian Gulf War.

Most of Gen. Norman Schwarzkopf's key planning officers were graduates of the School of Advanced Military Studies (SAMS) at Fort Leavenworth, Kansas. The school was established in 1983 by Brig. Gen. Huba Wass de Czege to teach top army officers the operational art of war. The strategy taught focuses on maneuverability, agility, and speed rather than the army's traditional tactics of superior firepower and frontal assaults. Colonel Boyd regularly gave his briefing on the importance of fast-cycle warfare. The school tried to foster a "bond of common knowledge and common interest" among the students, according to Col. Richard Sinnreich, a former SAMS director. "Huba's view, and mine, was that common orientation was not only valuable, but essential." This bonding "facilitated within the Army an ability to communicate tactical and operation constructs concisely and clearly," according to one former deputy director. In the Gulf War, every level, from division and corps command up to U.S. Central Command in Riyadh, had a planning group of three to five officers, most of whom were SAMS graduates.[26]

The basic lean and clean paradigm doesn't change much, yet it is well suited to the business world, which does. Successful adaptation will require constant re-education. This is also a crucial element of flexible manufacturing: a well-educated workforce, highly trained, yet continually retrained. New Japanese auto workers get 370 to 380 hours of training compared with 170 hours for European auto workers and 46 hours for U.S. auto workers.[27]

When times get tough, a flexible manufacturer does not immediately cut training. Reeling from slower demand and tougher U.S. competition, Honda Motors announced that it would trim production in the first quarter of 1993, the first time in ten years of operation. Rather than lay off 200 of the 5400 workers, Honda decided to use the extra time to intensify training. Honda workers receive training for jobs held by co-workers, as well as for "mundane" practices such as cleanliness and low-level maintenance. Harry Katz, professor of industrial relations at Cornell University, has pointed out, "It's hard to find American companies that are going the training route, but they must if they want the flexibility that will allow them to absorb the shocks of the economy and the market."[28]

After Dow Chemical set up its two-and-a-half-day Continuous Process Improvement Workshop, the number of winning projects in its annual Energy/Waste Reduction Always Pay Contest increased significantly. Training is an essential element of lean and clean management, increasing flexibility, promoting continuous improvement, and making layoffs a last step, not a first one.

Motorola is considered the "gold standard of corporate training" according to *Fortune*. It spent 3.6 percent of the payroll on training in 1992. The course catalog for Motorola University is the same size as most cities' yellow pages. William Wiggenhorn, president of Motorola U., says, "When you buy a piece of equipment, you set aside a percentage for maintenance. Shouldn't you do the same for people?"[29]

Motorola estimates that for every dollar spent on training, it gets thirty dollars in productivity gains within three years. There are few investments with such return. The company cuts costs by training workers to simplify processes, reduce cycle time, and cut waste. One team from purchasing was able to reduce the number of steps in handling a requisition from seventeen to six, and the hours it took from thirty to only three. How was the team able to do that? They had taken a two-day class at Motorola U called High Commitment, High Performance Team Training.

Training works because once a company has settled on the lean and clean approach, the same basic problem-solving techniques may be used endlessly: improving quality, reducing cycle time, performing life-cycle analyses, focusing on prevention, eliminating waste, fostering teamwork. Most of these techniques share similar principles. "The actions you might take to reduce waste paper," says materials manager Yakmalian, "might be very much the same as the steps to reduce cycle time." Once workers are trained to eliminate wasted time, training them to reduce wasted resources is much easier.

When someone is trained properly in applied systems thinking, such as reducing the cycle time of any process or product, that person gains a broad set of skills that map readily onto other systems problems. He or she gains an intuitive sense of how to tackle systemic problems.

> **Intuition is the pinnacle of systems thinking: the ability to see in an instant the solution to a whole problem. Intuition is fast-cycle thinking.**

Training workers to understand and improve only their own operations will perpetuate narrow, linear thinking. Training workers to understand and improve processes will develop intuition. That's why cross-training can be so powerful.

"The ability to learn faster than your competitors may be the only sustainable competitive advantage," says Arie De Geus, head of planning for Royal Dutch/Shell.[30] If so, then the only sustainable competitive advantage is training management and line workers in lean and clean techniques.

III AND IV. FASTER DECISION AND ACTION

Once a company improves its ability to observe the world and has begun to reorient its philosophy, it must complete the O-O-D-A loop and improve its ability to decide and act.

The ability to decide and act quickly depends on a company's having the fewest layers of management. Most big U.S. companies remain mired in bureaucracy, with top-heavy management systems.[31] Decision making should be delegated to those in the field wherever possible. The essence of flexible manufacturing is just such delegation—local responsibility and decentralized authority.

Delegation is essential to making the paradigm shift to lean and clean. People are not resistant to all kinds of change. They are resistant to change that is imposed on them, to change that they have no control over. Top management's job is to decide *what* kind of change is needed and to explain to everyone else *why* such change is necessary. But if employees are not brought into the process of determining *how* the change will occur, then change will be very slow indeed. This is particularly true for a company trying to become fast-cycle.

The fastest O-O-D-A loop comes when the line workers handle most elements of the loop, including decision and action. They should be responsible for most quality control, maintenance, and minor repairs of their machines. They should assist management with equipment purchase and layout. The ultimate responsibility, which requires the most trust but is the most crucial for success in fast-cycle companies, is to allow the line worker to stop the entire production line whenever a defect or mistake is found.

Common orientation and training help ensure that everyone in the company is working toward common goals. For Colonel Boyd, the common outlook will "simultaneously encourage subordinate initiative yet realize superior intent."[32] Common orientation helps top managers set goals and gives them the confidence needed to set workers free to meet them.

Line workers should be organized into self-managing teams, such as Quality Circles, that make suggestions and decisions about improving product and process. Teams can examine problems at a much deeper level than individuals, can generate ideas from a broader perspective, and can carry out a plan with a higher degree of sophistication. Individuals improve operations, but teams improve processes.

To Quality Circles, add Pollution Prevention Circles. Better still, combine the two and create Lean and Clean Circles. Once line workers are doing much of the work the supervisors used to do, fewer middle managers are needed. Those who remain act primarily horizontally within the organization, to help smash functional boundaries, to build networks.

Consider the lesson from nature's most successful and dynamic learning machine: the human brain. The human brain takes a systems approach to learning—it becomes smarter by creating new *interconnections* rather than by adding new brain cells. Similarly, a company becomes a learning organization only by creating new connections between management and labor and among its myriad departments. People and groups that never used to talk to one another now join cross-functional teams.

1. Cross-functional Teams

Probably nothing slows down a company or hurts quality more than the compartmentalization of functions. Separating the groups who design, manufacture, and sell has been seen as the road to efficiency. Instead, such an approach creates an us-versus-them mentality *within* the company. This worsens bureaucracy and fosters finger-pointing rather than problem solving. The company's interaction with the customer is reduced to one or two groups—sales and marketing—and even those may be forced to spend most of their time on internal problems.

Worse, compartmentalization creates sequential or linear problem solving. The design people come up with a great idea. They hand it off to engineering, which then hands it off to production, which may give it back to design or engineering because the product can't be built easily or cheaply. Or maybe production just goes ahead and makes it anyway.

The problem with linear problem solving is that a company's tasks are not *linear,* they are *systemic.* Fast-cycle means, for example, designing products that are easy to manufacture. Clean production, in mirror image, leads to products that are easy to take apart and recycle. Harvard Business School researchers studied twelve computer companies, dividing the companies' approaches to research and development into two groups: traditional (i.e., linear) and "system-focused."[33] The latter used cross-functional teams to solve problems largely ignored by the former, including manufacturability and productivity. The 1993 study concluded that the systems approach not only is faster and lower cost but accelerates the learning process:

> System-focused companies in our study were faster and more productive on individual projects and . . . the performance gap between traditional companies and system-focused ones actually increased over time.
>
> —Marco Iansiti, Harvard Business School

What steps can a company take to create a successful cross-functional team?

*Involve **everyone**.* A complete team encompasses *all* the O-O-D-A functions. With short lines of communication, the team's O-O-D-A loop will be fast. A cross-functional team maximizes *feedback*. The lean and clean design team for the Comstock building included architects; engineers for HVAC, electrical, communications, and structural systems; space planners; and interior designers. The project manager had both architectural and engineering training. As the chemical maker Fisher Scientific learned from its pollution prevention efforts, "the only effective approach" for their waste minimization is having "a full-spectrum multi-disciplinary team analyzing the problems."[34]

Everyone also means outsiders, including customers and suppliers. Only customers know how they actually use a product, what they like and dislike about it. Delivering high-quality, fast-cycle products or services requires suppliers that are high-quality and fast-cycle. That means you must help them meet your specifications.

A company must be imaginative in breaking down barriers. In designing the Taurus, Ford Motor Company realized that successful anticipation and adaptation may require novel design principles and diverse team members. Team Taurus brought in insurance companies to help design a car that would minimize the customer's repair costs when accidents occur. They used legal and safety advisers to help plan for trends in safety laws, rather than patch changes later on. The Taurus turned out to be not only one of the best-selling cars ever built but also one of the safest.[35]

*Create a genuine **team**.* Team members should not be part-time. Part-timers know that raises and promotions are determined by their own departments and are based on advancing in their narrow specialties—not on the team's success. Most internal team members should be full-time from the start. Their rewards and performance evaluation should be based primarily on *team* performance.

Team members should be trained together. Improving team dynamics is a special skill and requires special training. Motorola has found great success with using team training to reduce cycle time.

Team members should be working on the same floor, preferably in the same room. Physical separation leads to compartmentalization. As a senior executive at Honda has said, "We wish we could design, engineer, fabricate, and assemble the entire car in one large room, so that everyone involved could be in face-to-face contact with everyone else."[36]

Teams are essential to achieving fast-cycle pollution prevention (discussed more fully in the next chapter) according to George Larsen, the acting director of environmental management for Martin Marietta's Astronautics Group. The group found that the keys to success were getting "manufacturing and line employees to participate up front" and using "interdisciplinary task teams." In one case, "group synergy" allowed it "to change in three months what other companies need five years to do."[37]

Lean and clean requires teams.

2. Design for Resilience

Since almost any company can thrive in good times, the mark of enduring success is its resilience, its ability to thrive in bad times. Shigeo Shingo writes, "When repeated oil shocks inflicted serious damage on numerous companies, the extraordinary resiliency of the Toyota Production System brought it quickly to the attention of industrial circles around the world." After introducing a new model, American automobile plants require eleven to twelve *months* to get quality back to the level the plant had for the previous model. Using fast-cycle manufacturing, Japanese automobile plants return to normal quality in only six *weeks,* even though their quality standards are significantly greater than those of American manufacturers. Part of that resilience comes from having a faster product cycle, which means greater responsiveness and a faster learning process.[38]

The other essential element is how Japanese companies train their workers and harness their diverse experience and talents through suggestion systems and cross-functional teams. This diversity does not lead to confusion because the workers and the managers are steeped in a common orientation—quality, cycle time, and constant improvement—which helps achieve internal harmony.[39]

Harnessing diversity in a fast-cycle company creates resilience. It allows the company to master a variety of rapid and unique responses to any competitor, or any other danger. The most sophisticated learning machine ever developed by nature, the human brain, takes advantage of variety in a similar way. The developing brain attempts many neuronal connections, apparently not knowing which ones will prove successful. Those that reach and activate target cells survive. The others die.[40]

Ultimately, if a company becomes a learning organization it not only responds quickly to problems but can actually learn to anticipate them. Although this ability is rare in a big company, Motorola at least understands the point: "Anticipation is a religion at Motorola," noted *The Wall Street Journal,* "codified in the culture in a number of ways and designed to keep critical information flowing quickly to the top."[41] Motorola fosters anticipation through excellent observations, highly trained employees, and a culture that encourages dissent.

Observing, orienting, deciding, and acting lead to more observing and then through the whole cycle again. Endlessly repeating the O-O-D-A loop creates a flexible, resilient organization. The goal is to harness teamwork, training, and employee suggestions to lower costs, prevent problems, increase productivity, and operate at a higher tempo than the competition. The higher tempo allows a company to outmaneuver the competition, and, if it starts moving in the wrong direction, to reorient rapidly.

Anticipating problems and preventing them from occurring is the highest form of resilience. The fastest way to bounce back from a problem is to avoid it in the first place, and the best way to avoid surprises with either resource prices or environmental regulations is to prevent pollution.

8

LEAN AND CLEAN DESIGN
FOR FACTORY PRODUCTIVITY

We will not so lightly waste material simply because we can reclaim it. . . . The ideal is to have nothing to salvage.

—HENRY FORD, 1926

There are good reasons to protect the earth. . . . It's the safest and surest way to long-term profitability.

—PAUL ALLAIRE, Chairman and CEO, Xerox[1]

Waste is whatever does not add value to a product or service. Waste is not an inevitable result of production but rather a measure of its inefficiency. Recycling or reclaiming waste is commendable and useful, but neither eliminates the inefficiency in the process.

Having nothing to salvage—pollution prevention—is the ideal, because it avoids the problems surrounding pollution entirely; workers don't have to handle it, and managers don't have to worry about it or any regulations concerning it. More important, prevention invariably raises productivity because it forces a company to think about systematic process improvement. Pollution prevention drives a manufacturer to become more competitive and productive. Pollution prevention is the key to becoming lean and clean.

The productivity gain from pollution prevention is well documented, if not well known. Consider a 1992 study of chemical manufacturing by the nonprofit group INFORM that examined source reduction measures, only those measures that reduce or eliminate waste at the source. They do not include waste treatment, recycling, or any other measures to address wastes once they are created. The report concluded, "More than 95 percent of the source-reduction activities affecting product yield at 19 of the study plants increased production output. Ten plants reported an average increase in product yield of about 7 percent."[2]

At a Ciba-Geigy plant in New Jersey, the study found that two improvements in the dye-making process "made possible a 40% increase in yield, reduced iron waste by 100% and total organic carbon waste by 80% for the process, and resulted in annual cost savings of $740,000." Additional analysis showed that process improvement could increase the yield another 15 percent. And, as discussed in Chapter 2, a multidisciplinary team-based approach to process redesign allowed Fisher Scientific to identify twenty-eight improvements that cost a total of $79,000, increased production yield an average of 28 percent, and saved the company a total of $529,000 per year.

Another example of increasing productivity by decreasing waste comes from the authors of *Dynamic Manufacturing*, business professors Robert Hayes, Steven Wheelwright, and Kim Clark. They found that "reducing materials waste often improves productivity far beyond what one might expect from the material saving alone." Their study looked at total factor productivity (TFP), which is not merely the output per unit of labor but the product output as a function of all labor, capital, energy, and materials consumed in its production. Total factor productivity examines the overall efficiency of a process as opposed to the efficiency with which it uses any single factor, such as labor. The table below summarizes their findings in one plant:[3]

PLANT	AVERAGE WASTE RATE (%)	EFFECT ON TFP OF A 10% REDUCTION IN WASTE RATE
C-1	11.2	+1.2
C-2	12.4	+1.8
C-3	12.7	+2.0
C-4	9.3	+3.1
C-5	8.2	+0.8

The authors noted that "reducing waste . . . by 10 percent from its mean value (which by itself would reduce total manufacturing costs by only half of 1 percent) appears to have been accompanied by a 3 percent improvement in total factor productivity." This reveals the "powerful impact that reducing waste has on overall productivity."[4]

The goal is to minimize resource use and pollution in all three phases of a product's life cycle: during manufacturing, usage, and after the product's life has ended. Eliminating a pollutant entirely is better than trying to recover it from the environment. Here is the hierarchy for reducing pollution, starting with the best approach:[5]

1. *Elimination:* Total pollution prevention (example: switching from CFCs to an environmentally safer chemical)

2. *Efficiency:* Reduced use of a resource (example: using a more efficient electrical motor)

3. *Closed Loop Recycling:* In-house or in-plant recycling, allowing more efficient resource use while confining contamination and risk (example: reusing solvents on site)

4. *Open Loop Recycling:* Recovering spent resources from widely distributed products (example: recycling aluminum cans)

Pollution prevention generally has the greatest benefits. Although there are thousands of examples of successful pollution prevention programs, the efforts of Martin Marietta illustrate a variety of important points about how to reorient corporate culture.

SHIFTING MARTIN MARIETTA'S ENVIRONMENTAL PARADIGM

Martin Marietta's Astronautics Group employs 7200 people in Denver. They develop, test, and manufacture a variety of advanced technology systems for space and defense. Recent work has included launching defense payloads aboard the *Titan IV* rocket and mapping the enshrouded surface of Venus by NASA's Martin-built *Magellan* spacecraft. Here's what the group accomplished:

Working in teams, Martin Marietta discovered that by replacing existing chlorinated solvents with environmentally safe cleaners and processes they increased worker satisfaction, lowered maintenance and material costs, eliminated expensive disposal and even improved the quality of the cleaning performances.[6]

A 1993 article coauthored by George Larsen, acting director of environmental management at the Astronautics Group, presented a case study of the group's efforts at "shifting the environmental paradigm."[7] It noted, "Today, the Astronautics Group attitude is best described as proactive." The group believes "that pollution prevention—the process of identifying and eliminating the source of a potential pollution problem before it requires controls—is the best way a corporation can achieve and maintain regulatory compliance and, at the same time, reduce its future liabilities."

Where did the motivation for this paradigm shift come from? In 1985 traces of trichloroethylene were found in a shallow well at a municipal treatment plant for metropolitan Denver located only a few miles downstream from Martin Marietta's principal research and development and manufacturing plant. In 1986 the Colorado Department of Health levied a $1.2 million fine against the company, citing its mishandling of hazardous waste. The Department of Health and the EPA eventually approved a cleanup plan to alleviate both soil and groundwater contamination. Its projected cost to Martin Marietta: $60 million.

The Astronautics Group was determined never again to let environmental issues overtake the company, and in 1987 it formulated a pollution prevention strategy and program, which included investigation and implementation of new nonpolluting manufacturing processes, procedures, and equipment. The group took a lean and clean approach to operations and devised several projects. The first project targeted TCA (1, 1, 1 trichloroethane), the favored solvent for degreasing rocket components. It had been widely used for years, but international negotiations to protect the earth's ozone layer banned its use after 1995. The company examined six alternative aqueous cleaning systems and settled on Daraclean 282, in part because it can be cost-effectively reused after being decontaminated.

By 1993 TCA vapor degreasing was completely eliminated. The company reported that "our experience shows that the long-term economic benefits far outweigh the up-front costs." In other words, the life-cycle costs of the new solvent were far lower than those of TCA. The up-front costs included $70,000 in study and research plus $200,000 in implementation. The benefits include $50,000 yearly savings in material costs; $400,000 a year in avoided ozone depletion taxes through 1995; and a $150,000 yearly reduction in operating, maintenance, and waste disposal costs.

The elimination of TCA is also a good example of how systemic or process change invariably brings extra benefits. As the article noted, "The aqueous agent is biodegradable, easily recyclable, and has no known harmful environmental effects. It also has no known employee health risks and actually cleans aluminum surfaces better than the TCA it replaced."

A second project was aimed at eliminating CFC-113, another ozone-layer-destroying chemical, used in the cleaning of spacecraft components, which require high levels of cleanliness to ensure their operating reliability. For this project, the company put together a multidisciplinary integrated product team of engineering, operations, safety, and environmental management personnel. Their objective: Using a "systems approach," identify and evaluate alternative materials and cleaning process equipment that meet end-cleanliness requirements.

The replacement system chosen was an alcohol-based spray, in part because the company had a twenty-five-year history of success using alcohol in other cleaning operations. The company built on what it knew. The up-front costs of the change came to $1.3 million, with annual savings in material and disposal costs of $325,000. The change will pay for itself in four years.

A third project is eliminating toxic chemicals from hand-cleaning operations. Ten alternative cleaners were examined, and two were chosen for sixteen months of exhaustive testing. The tests examined the new cleaners' ability to remove six typical contaminants from aluminum panels before adhesive bonding and revealed that a citrus-based cleaner was superior to the existing cleaner, leaving less residue, which in turn resulted in higher adhesive bond strength.

The change generated no capital costs aside from study and research expenses of $350,000. Toxic emissions were reduced

by thousands of pounds a year. Yearly savings are estimated at
$250,000. The company commented, "Worker satisfaction has
been significantly enhanced by the switch to a citrus-based
cleaner." The orange-smelling liquid is gaining broader acceptance
than expected: "Employees say they prefer the citrus-based cleaner
because it's more efficient, successfully removing all contaminants
encountered to date."

The benefits of change have been enormous, but the study identified several barriers that had to be overcome in adopting a proactive
attitude:

- *I've always done it this way. . . . We've been using this solvent for
  thirty years.*
  Senior quality engineer Rob Musser noted, "People get so comfortable and used to dealing with a mature process, naturally they resist
  change."

- *If it ain't broke, don't fix it.*
  One leader in the prevention effort said, "As long as our operations
  are consistent with legal requirements and accepted industry-
  disposal practices, there is a tendency to believe mistakenly that
  we've done all we can."

- *Our mission is critical; the environment must take a backseat.*
  For workers whose products are used in the nation's defense, it is
  easy to assume that the importance of their products shields them
  from the burdens of environmental improvement.

- *We've already done all we can.*
  It's easy to lose momentum and rest on success once initial targets
  are met.

- *It will make my job more difficult.*
  Workers fear changes will increase labor, hurt performance, upset
  customary work practices, force organizational change, or require
  huge capital spending.

- *We'll have to sacrifice quality performance.*
  Even with extensive and favorable testing, "process changes were
  initially viewed as unreliable" and thus met with "deep resistance."
  Yet as the study reported, "The success of these projects proves that
  environmental alternatives often enhance, rather than sacrifice,
  quality."

In the course of its paradigm shift, the Astronautics Group found several keys to overcoming these barriers and building a permanent culture of pollution prevention:

1. *Ask provocative questions.*
"What if we transformed our old equipment to handle different materials?" Common practices must be challenged if unexpected win-win answers are to be found. One worker said he "never thought something squeezed from an orange or lemon peel would be superior to conventional chemical solvents."

2. *Create a climate of openness.*
The group found it was critical to get manufacturing and line employees to participate up front and to use interdisciplinary task teams. In the TCA case, group synergy allowed it "to change in three months what other companies need five years to do." This is fast-cycle pollution prevention.[8]

3. *Lead with technical language.*
Make clear that the environmentally beneficial change is subject to the same rigorous testing other processes are subjected to. Run pilot projects to demonstrate that the new approach works. Share innovation. Martin Marietta uses teleconferencing to link separate facilities and departments to promote the flow of knowledge and experience.

4. *Build skills, awareness, and an environmental ethic.*
Educate workers about the environment. Train them in pollution prevention. Survey them. A 1991 group survey showed that 63 percent of the workers agreed that "if there is any one group in our country that is responsible for damaging our environment, it is big business." Nearly 75 percent *disagreed* with the statement "Full compliance with all relevant federal, state, and local laws and regulations is enough." Yet, 91 percent agreed that "good environmental performance is good for the company's bottom line." Finally, 85 percent agreed that if "Martin Marietta demonstrates a stronger commitment to the environment, it will make it a better company to work for." A commitment to pollution prevention boosts morale, which in turn boosts the company's commitment to prevention.[9]

5. *Stay true to your goals.*
When the pollution prevention staff's projection came out ninety tons short of the corporate goal and they "saw no possibility of further

reductions," the environmental director, Bob McMullen, refused to accept it. He said, "I saw your numbers, but you're still responsible for meeting corporate goals. Go back to the drawing board." The staff met the challenge.

6. *Exercise leadership.*
In 1990 the company became a founding member of Colorado's Pollution Prevention Partnership, a nonprofit, voluntary alliance of government, businesses, and public-interest groups organized to pool information and resources to further the common goal of pollution prevention. The partnership, which includes companies such as Coors and Hewlett Packard, reduced TCA use 90 percent from 1988 to 1991, eliminating a million pounds of hazardous chemicals a year.

Martin Marietta discovered the lean and clean synthesis: a systems approach to pollution prevention—measuring waste, establishing goals, creating teams to reduce waste, implementing programs and process changes to reduce waste, and measuring progress. The case study concluded: "The formalized Total Quality Management (TQM) concepts of designing and building quality in everything the Group does . . . are remarkably compatible with the pollution prevention approach to environmental management." This provides support for the conclusion of a different study of pollution prevention in a large multinational firm: "The units that had strong TQM programs in place undertook more wide-ranging and effective pollution prevention efforts than divisions with less commitment to TQM."[10]

LEAN AND CLEAN DESIGN

The two highest goals for clean production—preventing pollution entirely and reducing pollution through increased resource efficiency—are both associated with higher productivity. They require *redesign* of both product and process, sometimes called design for environment. Lean and clean design results when design for the environment is coupled with design for manufacturability and other lean production techniques.

Lean and clean design not only cuts costs, increases productivity, and improves the environment but can also minimize business risk and health risks to employees. Braden Allenby, research director for technology and environment at AT&T, imagines a situation in which a competitor has designed a product that "not only contained no lead solder (where you had to use a lot), but was recyclable as well, so that a consumer trade-in program was initiated." He contends that the competitive impact would be profound, especially "if regulators felt compelled—or were required by statute—to mandate the lead-free technology."[11]

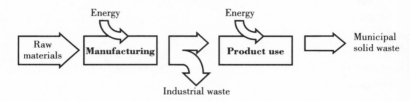

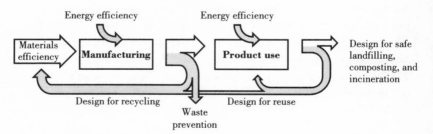

U.S. Congress, Office of Technology Assessment, *Green Products by Design: Choices for a Cleaner Environment*, OTA-E-541 (Washington, DC: U.S. Government Printing Office, October 1992).

HOW DESIGN AFFECTS MATERIALS FLOWS

In general, materials flow through the economy in one direction only—from raw materials toward eventual disposal as industrial or municipal waste (top). By making changes in a product's design, overall environmental impact can be reduced (bottom). Green design emphasizes efficient use of materials and energy, reduction of waste toxicity, and reuse and recycling of materials.[12]

The U.S. Office of Technology Assessment gave a pictorial view of green design in an October 1992 report, *Green Products by De-*

sign: Choices for a Cleaner Environment (see the figure on previous page). Green design requires knowledge about manufacturing, materials, packaging, and environmental impact. Also, customers know best how products are actually used and abused, what parts wear out or break down the fastest, and what kinds of waste are generated during use. Thus, lean and clean design requires talking to customers and cross-functional teams.

Consider the story of the UKettle, an electric teakettle developed by Polymer Solutions for Great British Kettles. Established in 1989, Polymer Solutions is a joint venture between Fitch Richardson Smith, one of the world's largest design firms, and GE Plastics, the world's largest manufacturer of engineered plastics or polymers. For more than a decade, GE Plastics had been pursuing ways to make its products easier to design for the environment. One of its senior executives, Uwe Wascher, a German, was convinced that U.S. legislation would eventually catch up to that of Germany, which was "far more proactive in conservation of resources and control of waste." The new joint venture would help them achieve design for the environment in a systematic fashion. It would offer clients a complete and seamless package from new product concept to commercialization, or, as GE Plastics put it, "from art to part."[13]

The UKettle is one of the first products that features design for disassembly: "Design that makes it economical to disassemble and recycle at the end of its service life."[14] On the surface, the most notable features of this large, white plastic kettle are the big, arching handle and the large, heavy coiled cord, which makes clear this is an electric kettle. But it is underneath the surface where the effects of the environmental design principles lie. These principles fall into two groups. The first are those whose benefits are primarily environmental. In the words of one of the designers:

- It is designed so that all product parts can be safely and easily separated and sorted at the disassembly facility.

- It is made of recyclable engineering thermoplastics that can be reground and manufactured into second- and third-generation products.

- It is more energy efficient than competitive models and boils water faster than stove-top models and microwaves.

The second group of principles are even more useful. They are *lean and clean design principles* since they are also essential to "design for manufacturability," a critical component of lean production. The easier a product is to manufacture, the faster the production cycle will be.

- The UKettle requires far less energy to manufacture because of design that eliminates unnecessary steps in manufacturing, assembly and disassembly.

- It incorporates tight tolerance design principles so that parts are consolidated and traditional fasteners are virtually eliminated.

- It includes two-way snap-fits located in the base for quick assembly and disassembly. Break points are clearly indicated on the snap-fits, which make separation easy for the disassembler.

It is not surprising that the UKettle, touted for ease of disassembly, is easy to manufacture. Assembly is the reverse of disassembly. Lean production requires ease of assembly; clean production requires efficient disassembly. This suggests that one broad unifying principle of industrial design for lean and clean production should be *simplicity,* which leads to a minimum number of parts and modular design. Modular components that can be upgraded individually simplify the introduction of constant incremental improvements, an essential element of fast-cycle manufacturing.

The faster the product cycle, the faster the learning process, and the more opportunities there are to learn from the feedback of experience.[15] As we learn more about environmental hazards and as regulations change in response, the company with the fastest product cycle and the most modular designs will be best able to adapt and remove the undesired material input or waste output. The design team for the UKettle included:

- Industrial designer

- Mechanical engineer

- Plastic applications specialist

- Tooling and molding engineer

- Graphic designer

- Marketing specialist

- Package designer

- Information specialist (who writes the user guide)

The UKettle project was not handed from one expert to the next. Everyone worked on it concurrently. The multidisciplinary design team achieved a remarkable result: "UKettle was brought to market in a record eight months, half the time usually needed for small appliances." The UKettle is not just environmentally friendly but fast-cycle. It is clean and lean.

Designing for Simplicity: Less *Is* More

Simplicity has many virtues. They are worth enumerating because complexity seems to be the rule in product design today. Complexity does not merely appear in the form of products with many parts that are difficult to manufacture and even more difficult to take apart and recycle. As one 1993 analysis noted, "Many manufacturers engage in a kind of features warfare. If one company offers twelve buttons, another must offer fifteen. The need for product differentiation produces another misery for consumers, which is that no products in any category ever work the same way."[16]

The authors of *Dynamic Manufacturing* give an example of the benefits of simplicity.[17] An equipment manufacturer redesigned a major hydraulic system to have a modular design with fewer parts. The results are shown in the table below.

Design for Simplicity in a Hydraulic System

	BEFORE	AFTER
Number of components	63	9
Time to market (months)	15	5
Labor productivity index	100	152
Product-quality indices		
Inspection costs	100	18
Customer rejects	100	47.5

Time to market dropped 67 percent: The product became fast-cycle. Inspections fell 72.0 percent, and customer rejects fell 52.5 percent: Quality jumped. Finally, labor productivity jumped 52.0 percent. Systematic product redesign, as with systematic process redesign, invariably raises productivity. Reducing the number of parts also saves money.

> **Nonexistent parts cost nothing to make, move, handle, orient, inventory, purchase, inspect, rework or service.**
>
> **—Ton Borsboom, industrial designer[18]**

A vending equipment maker found "significant cost reductions" from reducing the number of parts in one product from 241 to 101 and cutting assembly time from seventy-six minutes to seventeen, as related in a Harvard Business School case study. It is interesting that the cost savings "did not result from one primary goal of the project—using automatic assembly equipment which could not be fully justified because the final design was so simple to assemble." Further increases in production came from adding line workers to assemble the product manually.[19]

Automation can be a valuable tool, but if it is used to compensate for overly complex product and process design, it is merely covering up a systemic problem. It is speeding up an operation, not improving a process. It is wasting money that could go to engineers and line workers.

Another virtue of simple design is reduced environmental impact. Products become easier to disassemble and recycle. Consider what Whirlpool's German affiliate found when packaging engineers cut the number of materials in their appliance packaging from twenty to four. Disposal costs fell more than 50 percent, materials costs dropped, and performance improved.[20]

Design for simplicity can also mean "design for installability." Typical circuit breaker boxes can take hours to install because of their complexity. Fitch Richardson Smith reduced total parts by 76 percent for Square D Company's Trilliant Home Power System.

Trilliant has 6 basic parts compared with the 173 metal pieces used in previous generations. Designers asked how a new product could provide better services to its various users, including manufacturers, contractors, carpenters, electricians, painters, inspectors, and homeowners, and the resulting product "is manufactured more quickly, stores in less space through nesting, weighs less, has substantially fewer parts, installs in less time, is easier to drywall around, paint around, and visually inspect."[21] It also saves a contractor up to one hour in installation time. Designing simple products that are easy to service or maintain is *prevention engineering*, which achieves the goal of minimizing life-cycle costs.

Simplicity is also, increasingly, what the customer wants. Surveys have found that between a third and a half of Americans cannot work programmable equipment, such as VCRs that need to be programmed for time-delayed viewing. Former Matsushita President Akio Tanii is quoted as saying, "We began to realize there's a gap between what we've been making and what consumers want." To address this problem, Tanii sent teams of his employees, including product designers, to visit 1 million homes to find out how the company's products are being used and what Japanese consumers really want.[22]

A new school of "user-centered design" is evolving. According to one article, "A growing number of technologists think that the development process should be reversed," so that designers start with the customer to "reflect the people who use the machine rather than the engineers who want to show off new technology."[23] In other words, the design community may be starting to see the power of systems thinking: taking a "backwards" approach to problems and focusing on the end users. It's about time.

Design for simplicity is not simple. All end users must have a say: manufacturing line workers, the people who install the equipment, customers, and whoever deals with the product at the end of its lifetime. Yet even this cradle-to-grave approach is not enough.

DESIGN FOR REMANUFACTURE

The best companies are already thinking cradle to cradle, or cradle to reincarnation: remanufacturing a product rather than having the consumer just toss it away. According to a 1984 *Technology Review* article,

> Remanufacturing is an industrial process in which worn-out products are restored to like-new condition. . . . In a factory environment, a discarded product is completely disassembled. Usable parts are cleaned, refurbished, and put into inventory. Then the new product is reassembled from both old and, where necessary, new parts to produce a unit fully equivalent—and sometimes superior—in performance and expected lifetime to the original new product. In contrast, a repaired or rebuilt product normally retains its identity, and only those parts that have failed or are badly worn are replaced or serviced.[24]

The idea itself is decades old. Remanufacturing of automobiles on a significant scale dates back at least to 1929, when watchmaker Albert S. Holzwasser formed Arrow Automotive Industries Inc. to remanufacture automobile parts.

The typical remanufactured product keeps about 85 percent of the original components (by weight), creating significant environmental advantages: Remanufacturing is estimated to use one-fifth the energy and one-tenth the raw materials needed to make a product from scratch. And remanufacturing has another advantage common to many clean production measures: It is labor intensive. Junking a product and manufacturing a new one from scratch is capital-, resource-, and energy-intensive. With remanufacturing, workers are needed to disassemble a product; inspect it; repair, replace, or upgrade parts; and then reassemble it. Human brainpower and labor replace resource use.

Xerox is a leader in remanufacturing, offering the service for many parts in its copiers: electric motors, power supplies, photoreceptors, and aluminum drums. The company recycles some 1 million parts a year, for replacement components and in new equipment, which saves about $100 million a year. Yet these parts represent only *two* percent of the available total.[25]

Xerox recognizes that "significant financial opportunities exist which outweigh the environmental aspects of this business." As part of its overall Total Quality Environmental Management Plan,

Xerox has developed a strategy it calls "asset management—the management of products and inventory to minimize their environmental impact at all stages of the product life-cycle, particularly end-of-life." Because asset management is a complex process requiring "the integration of design, engineering, and re-manufacturing," Xerox created a cross-functional Asset Management Quality Improvement Team, led by Dick Morabito.

The team set up a program that has trained over 1000 design engineers in remanufacturing. Morabito notes, "Now they can work up front with the design teams so that the re-manufacturing capability is built into the product delivery process more explicitly and is ready at launch." Xerox engineers are taking a variety of steps in the design process to facilitate and improve remanufacturing. They are standardizing designs so more parts can be used in a wider variety of products. Remanufacturing lines were put in parallel with new product lines to match their level of quality. And, not surprisingly, Xerox has brought its suppliers into the design process in order to maximize opportunities for remanufacturing.

Chairman and CEO Paul Allaire asserts, "the environment is a business issue of strategic importance and Xerox must take the lead." In turn, Jim MacKenzie, Director of Corporate Environment, Health and Safety, notes that while the company's primary driving force is responding to customers and regulatory requirements, "we are not always creating designs to satisfy our customers now, but to anticipate the future customer requirements and business strategies." Xerox's goal is to be waste free by 1997, defined as greater than 90 percent reduction in waste. It is striving for "waste-free products manufactured in waste-free factories."

GE Plastics has examined the possibility of a simpler refrigerator that is easier to service and recycle. One key feature would be an electromechanical module designed for serviceability as well as disassembly. This module would contain temperature control, air distribution, and lighting functions. These elements have traditionally been scattered throughout the unit and can be difficult to separate for disassembly.

Modularity also makes continuous product improvement easier. Putting many controls in one module will make technology, microprocessors, and software far simpler. Getting the "latest model" would no longer mean purchasing a whole new product but rather a far less expensive new module. Such upgrading should help, not hurt, product sales. Hardware would become more like computer software, for which customer upgrades can happen every year or two. Customer loyalty to such software programs is tremendous, as companies such as Microsoft have found, maximizing the opportunity for sales.

Finally, designing products to be returned by the customer for remanufacturing has a competitive advantage:[26]

> Manufacturers can see first hand the types of failures that are actually occurring, and incorporate this information in future products, thereby improving quality.
>
> —E. Thomas Morehouse, Jr.
> "Design for Maintainability," 1992

The feedback of seeing products at the end of their life cycle will help the designers at the start of the next product life cycle.

THE CYCLE OF LIFE

Life-cycle analysis is essential to a successful business. Basing all decisions on initial or up-front cost, rather than long-term or life-cycle cost, commits a company to unnecessarily high future operating costs. Those costs will, in turn, undermine cash flow and reduce profits. They may eventually bankrupt the business.

Sometimes life-cycle analysis is simple. As noted in Chapter 7, Home Depot does not see its customers in terms of their average single-visit spending of $38 but instead calculates the life-cycle value of a customer at $25,000, which leads it to emphasize customer service and worker training.

Life-cycle analysis for full environmental costs can be far more complicated. One 1992 article on life-cycle cost management for pollution prevention defined the life-cycle cost of a chemical or material this way:[27]

Life-cycle cost = Cost of acquisition   Direct purchase, handling or transportation, record keeping, etc.

+ Cost of use   Direct costs of use, training and management, occupational liabilities, waste minimization efforts, etc.

+ Cost of disposal   Treatment, actual disposal, record keeping and management, etc.

+ Postdisposal cost   Long-term record keeping, potential legal liabilities, etc.

Many of the pollution prevention cases discussed in this book make use of a rudimentary life-cycle analysis. But very few comprehensive environmental life-cycle calculations have been made, at least in this country, because designing products and processes to minimize environmental impact is a relatively new concept. It is also not easy to estimate all these costs, especially for companies that have not been measuring environmental costs. Finally, a major obstacle to life-cycle costing is the compartmentalized nature of most companies. As Hewlett Packard, a member with Martin Marietta of Colorado's Pollution Prevention Partnership, noted in a 1993 partnership report:

> The concept of involving manufacturing and research and development in pollution prevention activities met with initial resistance, since waste handling had been traditionally handled by environmental departments. Distributing waste management costs directly to manufacturing cost centers facilitated cooperation, however, as it was realized that pollution prevention really can pay through process efficiency and reduced operational cost.[28]

In other words, a company's various departments may, in total, know all the relevant costs, but if all the people with the individual pieces of that knowledge are not connected, it may still fail. In the words of the ancient Sufi sage: "You think because you understand *one* you must understand *two* because one and one make two. But you must also understand *and.*"

If creating an "environmental department" disconnects the manufacturing department from the entire effect of its actions on the company's bottom line, the effort will fail. If manufacturing isn't made directly responsible for all costs that its pollution has incurred, it will naturally generate more pollution than your company would desire on strictly economic grounds. Manufacturing needs all the feedback of its actions—the positive and negative costs to the company—to make the best decisions.

ENERGY EFFICIENCY IN PRODUCTION: MOTOR SYSTEMS

Redesigning production processes to eliminate pollutants is the ideal approach, but redesigning for partial elimination is also valuable. A key target of opportunity is energy, not just because it is used so inefficiently in almost every industry but especially because utilities will often finance much if not most of the improvement.

Motors and motor systems offer the largest opportunity for energy savings in any industrial facility. Motors use a vast amount of energy—in the United States, about half of all electricity and almost 70 percent of industrial electricity. Yet motors are usually inefficient and oversized. A typical inefficient motor uses ten to twenty times its capital cost in electricity each year.[29] Thus high-efficiency motors, new control systems, and systematic process redesign afford tremendous opportunities for energy savings. These savings have been documented by a variety of companies.[30]

- A manufacturing plant replaced sixty-eight old motors in a buffing operation with high-efficiency motors, thirty-two of which were downsized 30 percent. Annual savings are $50,400 (630,000 kilowatt-hours per year). The payback was immediate because the utility covered the entire cost, $62,000. Even without the rebate, the payback would have been little more than a year.

- A major wastewater treatment facility was varying the water level in its aeration tanks *manually* to obtain the desired level of dissolved oxygen. A $300,000 upgrade put a new variable speed drive on each of the facility's eight 60- to 75-horsepower motors. A microcomputer now maintains the correct dissolved oxygen level. Annual savings are $60,000 (950,000 kilowatt-hours per year). The utility rebate was $102,000, so the net cost was $198,000—a payback of 3.3 years.

New high-efficiency motors can save twenty to one hundred dollars per horsepower over their lifetime compared with standard motors. Since factories are likely to have dozens of motors, the cumulative savings can be considerable. Nevertheless, because of their higher initial cost (10 to 30 percent more than standard motors), such motors make up only 20 percent of the national sales of motors larger than one horsepower. They constitute a mere 3 percent of existing motor stock, yet, when included as part of an entire motor system changeover, paybacks are often under two years.

The payback is even faster when the local utility provides a rebate for energy efficiency. About two dozen North American utilities offer such rebates, averaging about ten dollars per horsepower for high-efficiency motors. The savings achieved by such programs have been in the range of $0.001 to $0.005 per kilowatt-hour, which is exceedingly cost-effective.

When motors over ten horsepower fail, they are commonly rewound rather than replaced. Rewinding means stripping out the old windings and replacing the wire. Planning ahead and finding a distributor who carries energy-efficient motors creates another option: replacing the failed motor with an energy-efficient one. There are three reasons to do so. First, the local utility may offer a rebate for efficient motors. Second, rewinding often results in damage to the old motor, decreasing its efficiency and making a new energy-efficient motor even more cost-effective. Third, the damaged motor may well be oversized.

Motor systems often are oversized. Some leeway for the needs of peak demand may make sense, but when motors are excessively large, big savings will be possible by replacing them. Most motors achieve maximum efficiency at around 75 percent of maximum load and start to lose significant efficiency below 40 percent. Surveys suggest that about one-fifth of motors above five horsepower are running at or below 40 percent of rated load. Such oversized motors waste considerable energy and should be replaced for three reasons:

1. A smaller motor is less expensive.

2. Operating a more efficient motor at a more efficient load saves energy.

3. High-efficiency motors tend to maintain high efficiency over a broader range: While a standard motor might begin losing efficiency rapidly at 48 percent of full load, a high-efficiency motor might not drop off until 42 percent.

Downsizing is not the only option. Sometimes it is better to put controls on the old motor. Electronic speed controls (adjustable-speed drives) permit large motors to operate more efficiently at partial loads. These adjustable drives not only save energy but improve control over the entire production process. Microprocessors allow these drives to maintain more precise and accurate flow rates. Also, if the production process needs to be redesigned, adjustable drives provide the flexibility to operate the motors at a different speed without losing significant energy efficiency.

Always avoid the "one-size-fits-all" approach to motors. One manager of a major U.S. company proudly reported to Ron Perkins, now at the consulting firm Supersymmetry, how the company had "solved its motor efficiency problem" using adjustable speed drives. It saved 50 percent of its motor energy by putting drives on its motors, most of which had been running at partial load. Perkins did not want to burst the manager's bubble by explaining that if the company had first surveyed its motor systems and then bought smaller, correctly sized motors (and put drives on some of them), it could have saved 80 percent of motor energy. Again, the end-use approach—finding out what is needed before acting—is invariably the least-cost approach.

Moreover, even if a change reduces energy use and operating cost, no company likes to disrupt its operations. So for a change as disruptive as a motor upgrade, it is particularly important to get it right the first time.

Motors are part of an entire system, which includes the equipment the motor runs, such as a fan or pump, and the transmission or drivetrain, which transfers the motor's mechanical power to the equipment. After doing a complete motor inventory, examine some of the broader systems questions: Would the motors benefit from improved lubrication, maintenance, and tune-ups? Could process redesign reduce the motor's load? Is energy being wasted in transmission?

For example, about one-third of motor transmissions use belts, the most common being V-belts that rely on the friction between the belt and the pulley to transmit the torque. New belts—such as cogged V-belts, synchronous belts, and high-performance, thin, flat belts—are more efficient and last longer. The Inland Rome Lumber Company mill in Rome, Georgia, found that switching from V-belts to synchronous belts saved energy and reduced maintenance and replacement costs, providing a payback of under two years.[31]

For pumping fluids and gases, the savings from motor system redesign can be disproportionately large. For example, when you need only half the flow from a pump, you can theoretically save seven-eighths of its power. Motors often are sized for peak requirements that rarely occur. A pump might operate at 40 percent of its rated power 90 percent of the time and at 80 percent of its power 10 percent of the time. A smaller pump in parallel with the first could cover peak requirements, saving considerable energy while providing additional reliability because both sets of pumps would be unlikely to break at the same time.

It is equally important to investigate the 10 percent of the time the motor operates at 80 percent of load. Ask the "five whys" and other questions. Why does the water use surge? Why is there a surge at that time? Is the surge caused by poor system design that creates the bottleneck? Can process redesign eliminate the bottleneck? What is the cost of the redesign compared with the savings in energy use?

One principle of energy saving is: *Reduce peak loads.* Utilities try to reduce peak power demands to avoid costly excess generating capacity. Any company should reduce surge requirements for similar reasons. Lean and clean producers try to avoid surges because they reduce efficiency. As Taiichi Ohno wrote, "On a production line, fluctuations in product flow increase waste."[32]

One opportunity for motor system redesign occurs in any industrial process that requires both heating and cooling of the process stream. This straightforward change, called *pinch technology,* can track the heat flow, find the best configuration for process equipment and heat exchangers, and cut energy costs (and emissions) dramatically with rapid payback. Steuben Foods, Inc., a dairy products processing plant, did a pinch analysis with the help of the New York Power Authority. The resulting design changes, which required a heat exchanger, a heat pump, and variable-speed motors, were projected to save $290,000 a year on energy costs, a payback of only 1.4 years. Consider a pinch analysis for chemical plants, food processing plants, oil refineries, pulp and paper mills, or textile plants. Case studies of such process redesign in those industries have found savings from $31,000 to over $4 million with an average payback of 1.5 years.[33]

Systematic process redesign is not only the most powerful technique for reducing the energy consumed by motors but also the approach most likely to raise productivity. The Regal Fruit Co-op experience discussed in Chapter 2 is a good example. Process improvement and computer control provided energy savings that would have paid for themselves in three to four years but led to productivity and quality gains that exceeded the energy savings by a factor of ten.

SYSTEMATIC ENERGY SAVINGS: THE CASE OF SOUTHWIRE

The Southwire Company, a large manufacturer of copper rod, cable, and wire, had soaring energy costs in the early 1980s. Energy bills had reached 20 percent of overhead, up from 10 percent a decade earlier, and were rising 15 to 20 percent annually. Southwire's profit margins were dropping, and the company had to lay off 1000 workers.[34]

Top management decided to embrace a comprehensive approach to energy savings. It set a goal of 35 percent savings, where many in the industry thought 20 percent was the most that was attainable. Ultimately, Southwire saved 60 percent of its gas and 40 percent of its electricity per pound of rod, wire, and cable produced. To achieve these remarkable results, Southwire's energy management team used a combination of efficiency measures: new motors, efficient HVAC, and improved lighting, including new skylights in all its plants. The motor efficiency program was particularly systematic.

Every standard motor under 125 horsepower that fails is replaced with a high-efficiency motor rather than rewound. For larger motors, Southwire compares the costs and savings of rewinding versus buying new motors, and it buys a new motor if the net present value of the savings over five years exceeds the financing costs. That is, it chooses the option with the lower five-year life-cycle cost.

Southwire favors new motors because it has experienced reliability problems with rewound motors and is concerned about efficiency losses from the rewind process. Engineers check the equipment and tag motors that are clearly oversized; when these burn out they are replaced with correctly sized, efficient motors. The company stocks some new replacement motors on-site. It buys large numbers of new high-efficiency motors primarily from one supplier. Southwire expects that supplier to have high-efficiency motors in stock at all times.

Thus, Southwire's motor replacement policy is best described as proactive—a systems approach. Waiting for a motor to burn out before deciding what to do would be a poor policy. Like most manufacturers, Southwire is designed for continuous operation. Employees are paid at a piece rate (with profit sharing) and consequently don't want to shut equipment down. When a motor burns out, they want it replaced immediately with whatever is available. If the company had not determined in advance what it intended to do with a burnt-out motor, and if it did not have high-efficiency motors readily available, people would inevitably use whatever motor could be obtained fastest. Not only has the proactive policy enabled the company to ensure that its motor supplier has high-efficiency motors but Southwire's volume purchases have allowed it to negotiate a price for high-efficiency motors that is only 5 percent above the cost of standard-efficiency motors.

In one plant Southwire had very high electric bills during the utility's peak demand hours, and the energy manager wanted to save money by turning off equipment during those periods. Employees argued that they would lose $50 every hour the machines were shut down. The energy manager, however, had measured the costs of operating the machines during peak periods at $200 an hour. That measurement quickly changed the employees' minds. The plant now turns off seventy motors during peak periods.

What did Southwire's systems approach to energy mean for the company? From 1981 to 1988 Southwire estimates that it has saved $40 million in energy costs, which have decreased from 20 percent of overhead to under 13 percent. During a rough financial period, these savings were equal to *almost all* the company's profits. Energy efficiency turned around its profit margin and may well have saved the company—4000 jobs at ten Southwire plants. The potential economic and environmental benefits to a company from a systems approach to maintenance are equally great.

PREVENTIVE MAINTENANCE

How many of us do not regularly floss our teeth or visit the dentist? The effort, the cost, and the trauma are just too great. Yet anyone who knows somebody with a serious tooth or gum problem knows it is painful, expensive, and traumatic.

People tend to skimp on preventive maintenance. Preventive maintenance means spending money now for problems that either aren't yet serious or aren't there at all. It requires regularly changing the oil in your car even though the motor seems to be running fine. It requires sacrificing today for future benefits. It requires understanding all the life-cycle costs associated with any action, a rare understanding indeed.

The country as a whole skimps on preventive maintenance. A March 1993 *Scientific American* article, "Why America's Bridges Are Crumbling," began: "Inadequate maintenance has piled up a repair bill that will take decades to pay off. Indeed, the scope of the problem is only now becoming clear." New York City spends about $5 million a year on maintaining bridges and $400 million a year on rebuilding them and other capital expenses. A study by a consortium of universities found that by spending $50 million a year on maintenance, the city would need to spend only $100 million a year on capital expenses.[35]

The potential rewards to individuals and governments of preventive maintenance are enormous. The rewards for companies are equally large. Maintenance has been estimated to account for 15 to 40 percent of total operation expenditures. One survey found that for plants with an operating budget between $2.5 and $10.0 million, maintenance eats up at least $1.0 million. According to the Japan Institute of Plant Maintenance, "Total productive maintenance aims at maximizing equipment effectiveness with a total system of prevention throughout the entire life of the equipment. Involving everyone in all departments and at all levels, it motivates people for plant maintenance through small-group and voluntary activities."[36]

Total Productive Maintenance (TPM) is lean and clean thinking applied to maintenance. It has three key elements:

- *Preventive maintenance*: Scheduling regular maintenance to guarantee continuous, high-efficiency operation

- *Predictive maintenance*: Finding troubles before they exist and determining the life expectancy of parts so they can be replaced at the best time

- *Maintenance prevention*: Selecting or designing equipment that is easy to maintain and service, as well as improving existing equipment

What can TPM achieve? One Nissan plant attributed a $400,000 savings to TPM in 1986, equipment breakdowns went down by 80 percent, and the reject rate fell to 0.1 percent from 0.6 percent. Maintenance labor hours dropped 20 percent. Operators spend only about 3 percent of their time on TPM, ten to fifteen minutes at the start of every day. Payback times are very short. Dai Nippon, one of the world's largest printing companies, spent $2.1 million putting TPM in place. It has reported paying for that investment within six months and saving $5.5 million in under three years. This high return on investment is typical of any systematic focus on prevention, such as pollution prevention and defect prevention.[37]

Total productive maintenance tries to achieve "zero breakdowns" through design, diagnosis, and maintenance. Only the workers can achieve that goal.

Until we believe that the expert in any particular job is most often the person performing it, we shall forever limit the potential of that person. . . . Nobody knows more about how to operate a machine, maximize its output, improve its quality, optimize the material flow, and keep it operating efficiently than do the machine operators and maintenance people responsible for it.

**—Rene McPherson,
Stanford Graduate School of Business[38]**

Employees welcome TPM, since it recognizes their importance and gives them more control over their work. One management consultant has said, "TPM is well received by the maintenance department because it permits them to do more technical work." Workers are as frustrated by equipment breakdown as anyone else, more so if their pay is linked to their work rate or if the company has profit sharing. Everyone in the plant must be made to understand the terrible cost of equipment downtime. In the automotive industry, "each minute of downtime can cost a company $1,200."[39]

One U.S. chemical plant embracing proactive maintenance made clear to everyone that "every 1 percent increase in uptime translates as a $500,000 increase in plant profit." Over a five-year period the plant increased uptime from under 50 percent to more than 70 percent while it reduced the maintenance budget to $18 million, from $21 million.[40]

Training will boost worker acceptance of TPM because employees will see it as improving their skills. In his book *Kaizen,* Masaaki Imai discusses the TPM training program adopted by Topy Industries' Ayase Works, which manufactures automobile wheels with 660 employees and 800 machines. Ayase Works provided in-house instruction for seventy foremen and other leaders in "lubrication, how to tighten nuts and bolts, basic electricity, hydraulics and pneumatics, and drive mechanisms." Four hours were devoted to each of these topics. Trainees learned, for instance, why applying too much oil may cause a machine to overheat. The trainees then trained the workers in these maintenance techniques.[41]

As workers gained knowledge, they identified numerous lubricating spots that had never been noticed before. They tightened a total of 240,000 nuts and bolts in the plant and then marked them with a line of white paint on both the nut and bolt. The paint stripe means that a quick glance can now tell whether a bolt is properly tightened.

Three years after adopting TPM in 1980, Ayase won the Distinguished Plant Award from the Japanese Institute of Plant Maintenance. Ayase had cut the number of machinery breakdowns that forced the production line to stop three minutes or more from 1000 per month to only 200. Labor productivity is up 32 percent, tool replacement time is down more than 50 percent, the equipment operating ratio is up 11 percent, the cost of defective parts is down 55 percent, and the inventory-turnover ratio is up 50 percent. Again, a systems approach invariably raises productivity. It also has numerous unexpected side benefits.

Imai notes that "morale is higher." Workers have "a much stronger attachment to the equipment with which they work." The maintenance crew has a more skilled and valuable job performing sophisticated maintenance tasks, diagnosing equipment, and training operators to do simple maintenance. Since the plant is cleaner and more efficient, the salespeople are now "eager to bring customers to the plant and employ plant tours as a marketing tool."

Total productive maintenance is effective because it is a systems approach: It focuses on prevention and on reducing life-cycle costs. As noted in Chapter 2, however, in most companies rewards are few for preventing bad things from happening. If you stop a crack in a wall from bursting and save the city of Chicago hundreds of millions of dollars, no one will ever know because the damage never occurred. How can you be properly rewarded for spending time and money today to avoid far larger costs tomorrow?

The solution is to use the "Golden Tool" of systems thinking: *feedback*. Companies will have to go beyond the suggestion box and institutionalize rewards for prevention. Prevention must become a basic part of everyone's job. As Imai says, "Top management must design a system that recognizes and rewards everyone's ability and responsibility for TPM." Reinforcing feedback in turn requires *measuring* performance. Possible performance indicators include:

- Total cost of the maintenance department per month

- Maintenance cost per unit of production

- Machine availability (the percentage out of service for maintenance and repair as a function of total time)

- Utilization (the percentage of equipment use time in relationship to total available time)

- Number of predictive maintenance corrections per month

- Electrical usage per month

- Total value of spare parts inventory

- Total capital costs for machine replacement per year[42]

Once a company measures each of these factors, it can reward the group that performs the best in each area. Ultimately, it should be able to determine exactly how much a group's exceptional performance is worth and reward its members proportionally. As each group develops historical data on its equipment repairs, downtime, labor, material, and energy costs, it will be better able to do life-cycle costing and determine when replacement is more cost-effective than continued maintenance.

Doug DeVries, former purchasing manager of Hyde Tools, says that life-cycle costing is really a matter of "putting your best heads together to identify all costs of ownership." He notes that the lowest purchase price sometimes leads to the highest total life-cycle cost. For purchases based on life-cycle cost, especially large capital expenses, DeVries recommends carefully estimating returns and then tracking real costs and savings. "Whenever I did life-cycle costing," he says, "I offered the accountants double the prime-rate return over the machine's lifetime. If I couldn't confidently predict a double-prime profit, I didn't propose the purchase." In the late 1980s and early 1990s, that meant DeVries was promising a 14 percent return on investment, which required him to do a seven-year life-cycle analysis.[43]

Total productive maintenance reduces equipment failure, setup and adjustment time, idling and minor stoppages, lower speeds, defects in process, and drops in yield, and it directly reduces manufacturing cycle time. It also has myriad environmental benefits. It reduces scrap and oil leakage, a major environmental problem at many plants. (At Ayase Works, oil leakage dropped from 16,000 to 3000 liters per month.) Energy consumption is reduced because proper maintenance reduces friction as well as wear and tear on the machines. Energy is also saved because the machines are more continuously running near maximum efficiency rather than wasting energy constantly starting or stopping. Companies have found energy savings as high as 30 percent.

Total productive maintenance reduces wasted time and wasted resources. It is both lean and clean. To pursue TPM, a good first step is to join the American Institute for Total Productive Maintenance, which has a monthly newsletter with useful articles and the names and phone numbers of TPM experts. And since benchmarking is one of the best ways to start any process of change, the institute schedules regular study tours of plants that use TPM. Start a library of TPM books. Several good ones are published by Productivity Press, including *Introduction to TPM* by Seiichi Nakajima and *TPM for America* by Herbert and Norma Steinbacher.

Recycling: Baxter Healthcare and Zero Waste

Recycling of materials is fundamental to lean and clean production. Many opportunities exist for waste recycling, both inside the company (closed loop) and outside (open loop).

For example, about 80 percent of manufacturing companies use solvents, but only a small fraction of those solvents are being recycled. By collecting spent solvents and boiling them in a simple batch still, different solvents can be recovered by virtue of their different boiling points. Haworth, Inc., a furniture maker, once used thirty gallons a day of organic solvents for cleaning. The total cost of waste management of its solvents came to $9000 a year. Replacing the spent solvents cost $30,000 a year. Haworth, on learning of the solvent recovery option, installed two simple batch stills, which recovered 75 percent and 90 percent of the solvents. The solvent that is recovered is very pure. The savings in waste disposal and solvent replacement paid for the recovery equipment in 1.0 year for one solvent and 1.5 years for the other.[44]

These and other technical opportunities for closed-loop recycling of hazardous waste have been written up as case studies. Your local EPA office should be able to help you determine if your waste products can be recycled by your own company or used by another company. Many of the best opportunities for recycling, however, will come from employees, as Baxter Healthcare learned.

Baxter Healthcare Corporation, I.V. Systems Division, is a manufacturer and packager of sterile solutions for intravenous use. It achieved remarkable recycling efficiency by focusing quality circles on waste reduction and recognizing employees who contributed to the company's waste reduction goals. It reoriented its workforce and reinforced the new direction with positive feedback.[45]

Baxter committed to constant, incremental improvement in reducing the plastic scrap generated in packaging. After the first goal, set in 1982, was met, lower goals for scrap generation were set, and so on. Recycled plastic is either returned to Baxter's manufacturing process or sold back to its supplier for reuse, saving Baxter $9 million from 1982 to 1992. Baxter now generates hardly any scrap—a mere *one eighth of one percent* (1200 parts per million), which is 90 percent lower than the company's scrap rate in 1982. As plant engineer John Carter reports, the recycling program "involved hundreds of projects and practically everybody in the plant."

In 1991 Baxter began to deal with the amount of its waste going to the local landfill and set a goal of reducing it by 43 percent in 1992. The company achieved a 51 percent reduction, or 3,036,000 pounds. The 1993 goal was to cut this by another 25 percent and practically eliminate waste to the landfill by the end of 1995, according to Carter.

Baxter recycles scrap plastic, cardboard, computer and white paper, aluminum cans, wooden pallets, fifty-five-gallon drums, scrap metal, and obsolete equipment. Baxter has eliminated Freon (CFC) as a cleaning agent. It has dramatically reduced lab chemical waste by installing a computer inventory system. In 1989 Baxter generated almost 12,000 pounds of waste oil. That has been completely eliminated by improved filtering, maintenance procedures, and reuse, which save the company $10,000 a year. No measure is too small. Colored legal pads, for example, have been replaced with white pads so that the paper can be recycled.

The profits generated from Baxter's recycling efforts, including sales of recycled material, savings from reduced new material purchases, and avoided disposal and landfill costs, now come to $1.7 million a year. By reducing waste and working with the state, the company has been reclassified as a small-quantity generator, greatly reducing its regulatory burden.

Baxter also has had a strong energy-efficiency effort since the early 1980s, including improved lighting, insulation, computerized energy management systems, and process changes. It is a member of EPA's "Green Lights" program. For a total capital investment of $4 million over several years, Baxter has reduced the amount of energy needed to produce its product by 48 percent, saving $3 million *a year*. Carter says, "One thing I always like to point out is that this just keeps going on as long as you keep your program going. It just keeps adding up."

Baxter has made these remarkable achievements with what it calls the Quality Leadership Process. This process helps employees learn teamwork and solve problems through teamwork. Baxter has different "quality working teams" dedicated to oil consumption, computer paper and aluminum cans, cardboard, cafeteria Styrofoam, plastic sugar/salt bags, wet polyvinyl chloride scrap, and wood pallets. The teams, consisting of six to twelve employees, meet regularly. Employees make suggestions and implement them through this system. With 2500 employees, Baxter receives 4000 suggestions a year. It has used 79 percent of them. All suggestions must be answered by the appropriate person within ten days.

Baxter Healthcare is a powerful example of how lean and clean thinking can produce tremendous savings from recycling and energy efficiency, made possible by tapping the brainpower of employees in a systematic fashion.

SYSTEMATIC RESOURCE REUSE: CASCADING

Cascading is a key approach to resource recycling. *Cascading* is using a resource repeatedly in different forms, a quintessential systems approach. Through intelligent process design, cascading can save a considerable amount of money while reducing pollution. Consider water. As the original resource (e.g., pure water) is altered or degraded by use, it is used for different purposes, often where lower quality (e.g., dirty water) is acceptable.

Cascading is described in a remarkable case study by the Institute for Local Self-reliance.[46] Buckeye Cellulose Corporation designed its Flint River wood pulp plant according to high environmental standards. The plant produces 270,000 metric tons of 85 percent brightness pulp every year for making diapers, while meeting strict EPA pollution standards. The goal of the pulping process is removing lignin (a polymer component of wood) from the cellulose fiber. The more lignin removed, the whiter the pulp. The removal is done in three basic steps: digesting, oxygen delignification, and bleaching. The pulp is washed with water after each purification step. A variety of cascading measures is used.

The washing water flows countercurrent to the pulp flow, which promotes recovery of the lignin and chemicals while reducing losses to the sewer. The white water removed from the pulp in pulp drying is used for wash water in bleaching. Filtered water from the last stage of bleaching is reused as wash water in the preceding stage. These steps minimize both the amount of wastewater used in bleaching and the amount that must be sent to waste treatment.

At similar mills, water is typically used to transport rejects (knots and long fiber bundles) to a press, where the fiber is removed for transport to a landfill. The removed water has high biological oxygen demand (BOD, which is harmful to the environment and common in food-processing wastes and sewage) and is usually sent to the wastewater treatment facility, an expensive process. In contrast, the Flint River pulp plant has a closed-loop system for refining and recycling the rejects, which eliminates the high BOD from the wastewater treatment system and also reduces the quantity of fibrous material sent to landfill. The case study notes,

> This strategic use of process water was carefully designed to allow for the greatest work to be derived from the least water. This was accomplished through strategically routing the water through processes requiring the purest water to those requiring successively less pure water.

While the best similar plants had been discharging about 6.00 kilograms BOD per 1000 kilograms of finished pulp produced, the EPA set a standard for new plants of 5.50 kilograms. In 1982 the Flint River plant averaged 1.15 kilograms.

Buckeye's corporate philosophy includes improving cost-effectiveness by recycling and resource minimization. It has an incentive award program for workers. Every employee is a member of a cost improvement program team. Each team sets goals for the team as a whole as well as for individual members (based on the plant goal). The plant makes annual awards to recognize outstanding contributions.

Cogeneration

Just as water can be cascaded, so can energy. *Cogeneration* is the making of power and heat together by the end user. Since most companies need these two forms of energy, manufacturing and service companies of all sizes can achieve great savings with cogeneration. The idea of "salvaging" heat rather than discarding it is an old one. Naturally, Henry Ford himself made use of it:[47]

Steam is required for heating the wood-drying kilns at five pounds per square inch pressure. Steam at 225 pounds pressure suitable for operating turbines can be produced at only 10 percent greater cost than that for the heating pressure. Thus, by developing steam in the power-house boilers at 225 pounds per square inch, passing it through turbines and "bleeding" low pressure heating steam from the turbines after a part of its available energy has been obtained, the steam is practically serving a double duty—supplying both power and heat.

—Henry Ford, 1924

Compact, simple cogeneration systems now exist. An internal combustion engine linked to a conventional generator creates the electricity; the engine water heat, normally wasted, provides the hot water. Combining the cogeneration unit with an electric heat pump also provides cooling. A small cogeneration unit can burn natural gas to generate electricity about as efficiently as a large utility power plant.

Any business that uses both electricity and hot water should consider a cogeneration system: restaurants, Laundromats, health care facilities, athletic clubs, motels, hotels, and schools. The savings can be so huge that the cogeneration supplier may agree to finance the sale through a shared energy sales agreement. For example, in 1987 GoldenBear Cogen Inc. provided a 7000-square-foot, 300-seat restaurant with a comprehensive cogeneration system, including:

• A sixty-kilowatt cogeneration module for electricity

• Two absorption chillers powered by hot water, replacing electric air-conditioning equipment

• A gas-fired hot water boiler to supplement engine heat recovery and large enough to handle the restaurant's entire thermal load if the cogeneration unit goes down

The electricity backup is achieved by tying the restaurant into the power grid. A microprocessor-based control system operates the system and allows remote monitoring of the cogeneration and HVAC system. GoldenBear Cogen owns and operates the equipment. First-year savings were $30,000. GoldenBear received $25,000 of this, providing a 4.4-year payback for the $120,000 system. The restaurant received $5000 of the savings as well as $10,000 more largely from the improved HVAC.[48]

When switching to cogeneration, rethink all your energy-saving measures. Such a systematic approach can save considerable energy and money while reducing pollution. Moreover, by bringing the energy supply inside a business with direct microprocessor control, cogeneration also maximizes flexibility. It can thus be an important element of lean and clean production.

How to Cascade Paper

Paper can also be cascaded. Buy recycled paper where practical, use paper more completely, and, ultimately, recycle. Several steps can be taken to use paper more completely:

- Internal company documents should be two-sided.

- Reuse the backs of old single-sided copies for memos and scratch paper.

- Designate someone to research the possibilities of standardizing the company's paper purchases.

- Try to limit the use of coated or colored paper.

- Set up a general memo-posting area or use electronic mail instead of giving many employees identical memos.

Once a company can no longer extract any value from paper use *inside,* it can recycle paper *outside.* Ask a local recycler to supply separate bins for white paper, computer paper, and mixed paper. Remove wastepaper baskets from copy machine areas and replace them with boxes marked with "recycle" symbols.

Global Turnkey Systems (GTS) is a $10 million computer system customizer in New Jersey with ninety employees. Worried about the recession, GTS wanted to trim costs in 1990. It embraced a comprehensive program of recycling (with some energy-saving efforts). Employees collect and sell eighteen tons of used paper a year, creating revenues of up to $150 a month. Recycled aluminum cans and glass are sold for about $35 monthly. Less trash has reduced garbage-hauling bills to $800 a month from $1500. A heating and cooling analysis (see Chapter 5) cut $2000 a month from the utility bill. High-efficiency faucets and toilets trimmed water bills almost 40 percent, saving $40 on the quarterly water bill, despite an increase in water rates. In addition to the considerable savings, GTS found a new market niche and now stages well-attended seminars that teach local companies how to save money while saving the environment.[49]

Three general principles for institutionalizing change are important for making any cascading and recycling program a success:

1. *Involve top management from the beginning.* At GTS the chief financial officer heads the ten-person volunteer cost-paring committee. If management won't recycle and won't do little things such as using two-sided paper, employees won't either.

2. *Involve everyone.* Create a detailed orientation program on recycling efforts. Fold the recycling program into an overall employee suggestion system. Involve maintenance workers and consult them for their own wise suggestions. Give all employees a separate desk folder or receptacle for recyclable papers.

3. *Give rapid, positive feedback.* When employees see directly that their actions produce results, they will be more excited about participating. Post monthly savings figures and graphs. Some of the savings could be used for employee incentive programs, such as parties and Earth Day bonuses, or donations to environmental groups or local municipal projects.

GETTING STARTED ON LEAN AND CLEAN PRODUCTION

Making the paradigm shift to lean and clean production starts by accepting the simple idea that pollution is not an inevitable result of business operations but rather a measure of its inefficiency. Once it is committed to lean and clean, a company's first step should probably be a comprehensive environmental audit. Such an audit will measure and analyze performance in environmentally related aspects of health, safety, and product stewardship. One of the best guides is the *Environmental Self-assessment Program,* available from the Global Environmental Management Initiative in Washington, DC.[50] It is a detailed, step-by-step guide for assessing a company's environmental performance.

The next step is to accept the idea that prevention is the approach to pollution with the highest return on investment and the greatest impact on productivity. Skeptical managers should go on benchmarking trips to companies that have demonstrated the best environmental practices. A focus on prevention requires intelligent design or redesign of products and processes, which in turn requires a cross-functional team-based approach and employees trained in lean and clean systems thinking. Prevention teams should do pilot projects and measure results. Top management should then set up prevention teams and goals for the entire company, such as a 33 percent reduction in energy use and waste per dollar of goods produced over three years. The most successful teams should be rewarded, and the goals repeated, with modification, endlessly.

9

THE FUTURE IS
LEAN AND CLEAN

Increasingly, the focus is shifting away from combatting the damage
caused by pollution and the disposal of industrial waste to the prevention of pollution during the production process.
—"The Commerzbank Report on German Business and Finance,"
The Wall Street Journal, October 22, 1993

The best way to predict the future is to invent it.
—ALAN KAY, Apple Computer[1]

The trend toward pollution prevention is one of the fundamental
shifts taking place in businesses throughout the world. A 1993
Commerce Department report on the global environmental export
industry came to the same conclusion as the German Commerzbank: "The environmental technologies industry is shifting from
technologies for pollution control, clean-up, and remediation (end-
of-pipe technologies) to those that prevent pollution from occurring
(clean production)."[2]

This book has been focused on the specific steps that businesses
need to take to increase profits and productivity by preventing pollution. A number of recent books and studies aimed more at policy
makers and opinion leaders have identified pollution prevention as
a fundamental national and global technology trend, including my
1992 book, *The Once and Future Superpower: How to Restore America's Economic, Energy, and Environmental Security;* and two 1994
publications, *Green Gold,* by Curtis Moore and Alan Miller, and *In-*

dustry, Technology, and the Environment by the Congressional Office of Technology Assessment.[3]

Even the World Bank, which over the past decades has financed so much traditional, resource-inefficient development, understands the shift. A 1993 World Bank Discussion Paper on Asia stated,

> The most financially viable environment-related investments are those that are good for both economic development and the environment, including energy conservation, waste minimization in industry (as opposed to end-of-pipe investments), recycling in the urban sector, fuel efficiency in the transport sector, soil conservation, and sustainable forestry.[4]

The future is lean and clean, and the reasons are clear. All the nations of the world are under assault from industrial pollution, and the best companies in the industrialized nations are learning the basic ideas behind lean and clean: Pollution is a measure of inefficiency, and prevention invariably increases profits and productivity.

The developing nations in particular are witnessing unprecedented population growth, urbanization, and industrialization. Resource inefficiency and environmental degradation are very real limitations on their attempts to raise the living standards of their people, especially since most of the developing nations of the world do not have the abundance of resources America was endowed with. The World Bank estimates that Asian countries alone will need to spend about $38 billion per year by the year 2000 on clean technologies to achieve greater sustainability. By the year 2000 the global market for environmental services and technologies, including energy efficiency and cleaner energy supply options, such as renewable energy, is projected to be as large as $600 billion. The resource, environmental, and capital constraints on the developing world guarantee that a rich export market will be available to the nation that is the world leader in developing clean technologies.[5]

The Japanese government is betting on clean technologies with heavy investments in energy-efficient and renewable energy technologies. In particular, it is vigorously pursuing the Asian environmental market through the Green Aid Plan, which is designed to help Asian countries fund projects aimed at prevention of water pollution, prevention of air pollution, treatment of waste, recycling, energy conservation, and alternative energy sources. In 1993 Japan quadrupled funding for the Green Aid Plan to $120 million.[6]

According to the Office of Technology Assessment, "Germany appears to be moving toward greater emphasis on pollution prevention." German regulations are increasingly pushing industry toward prevention, recycling, and life-cycle analysis. These and similar proposed or pending regulations throughout Western Europe have implications for U.S companies, as noted in a 1993 report prepared for the Saturn Corporation by the University of Tennessee Center for Clean Products and Clean Technologies: "European auto manufacturers are the current world leaders in car recycling and the use of life-cycle assessment to design environmentally superior cars."[7] As one German company has demonstrated, an environmentally proactive company that invents the future can achieve significant competitive advantage; a case study by Strategic Environmental Associates, a management consulting firm, described how:[8]

HENKEL ($7.6 BILLION SALES) ANTICIPATES AND PROFITS FROM INDUSTRY RESTRUCTURING

In the early 1970s, Henkel, a German detergent and chemical company, was the market leader in primary phosphate production and phosphate-based detergents. The company realized that its business was causing an environmental problem. Before any signs of regulatory action, Henkel developed a zeolite-based alternative, and in 1982 introduced its first phosphate free detergent.

By 1989, all of its products were phosphate free. Henkel's leadership served as a catalyst of German legislation, which in the mid 1980s mandated that all detergents be phosphate free. By that point, Henkel controlled 70% of European zeolite production and the market share for its leading brand of detergent had risen from 16% to 23% in Europe.

Henkel had successfully lead the restructuring of its market and influenced legislation to create a new industry standard from which its products were uniquely well-positioned to benefit.

The country most focused on prevention is the Netherlands. It spends about $500 million a year on environmental technologies, equivalent on a per capita basis to $9 billion in the United States, which far exceeds our environmental technology spending. More than one-third of this Dutch spending advances pollution prevention. The Dutch also use the tax code to promote clean technologies.[9]

The essential point of this discussion is that individual companies and nations alike should factor a shift toward prevention into their strategic planning.

FUTURE SCENARIOS: SUSTAINABILITY OR BUST

In January 1991 the Business Environment section of Group Planning for Shell International published a report entitled "Global Scenarios for the Energy Industry: Challenge and Response." These scenarios are worth examining because Shell is a major energy company with broad expertise in resource and environmental issues, and, as discussed in Chapter 7, Shell has been extremely successful at strategic planning, anticipating the effects of the oil crises of the 1970s and outperforming the competition as a result.[10]

In planning for these scenarios, Shell consulted with a variety of outside experts and examined three interrelated areas: sweeping geopolitical changes (including the collapse of Eastern Europe and the Soviet Union), problems in the international economy (such as trade frictions), and growing concern on individual, national, and international levels about environmental problems (such as resource depletion and global warming).

Shell's planners noted that the "consensus in favor of multilateral solutions, previously held together by a dominant United States, is in doubt. Environmental concerns can be divisive . . . but may also be an important force for cooperation." The developments in geopolitics, economics, and environment, they continued, "are all important signs of structural change in a global system under stress. However, the direction of change is not at all clear." The figure on page 171 illustrates "two possible directions—two alternative interpretations of the present signs of change" seen by the planners.[11]

In the Global Mercantilism scenario, the principal challenge is the weakness and instability of current international economic and

TWO ALTERNATIVE INTERPRETATIONS OF THE PRESENT SIGNS OF CHANGE

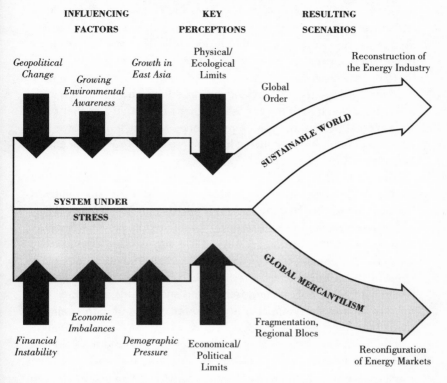

Source: A. Kahane, "Global Scenarios for the Energy Industry: Challenge and Response," Shell International Petroleum Company Ltd, England, 1991, p. 4.

political systems. The response is fragmentation. The world divides into regional trade blocs and is characterized by adversarial trade and financial instability. A "piecemeal" approach is taken to the environment.

In the Sustainable World scenario, the primary challenge turns out to be how to deal with common problems, especially ecological problems, such as global warming. Here the response is cohesion (as well as coercion) and a broadening of international systems. In the sustainable world, "politicians compete to be seen as the most 'green' " and "Environmental investments inspire invention and innovation, leading to profitable new economic activities." One of the results is a "resilient economic system."

Shell's planners noted that the purpose of the scenarios "is to sensitize us to recognize signals of possible changes in the world—which will probably include elements of both scenarios—and to enable us to respond quickly and appropriately. They are less reassuring than conventional forecasts, but more challenging—and therefore more useful."

Many reasons to reorient and become lean and clean have already been discussed: lower costs, higher productivity, higher profits, reduced pollution, greater resilience. To that long list we can now add another item: ensuring that the Sustainable World scenario becomes reality.

The world of Global Mercantilism is not a pleasant one. Living standards or quality of life may not improve for most Americans or for most people on the planet. And if any one of the serious projections of environmental and resource problems proves true, such as significant global warming, quality of life may decline dramatically for many. Of course, even in this scenario an individual company will still have to become lean and clean because competition between companies will remain fierce, and successful companies will need to pursue every opportunity to lower costs and gain competitive advantage.

The Royal Dutch/Shell planners are not saying how the world *will* be or even how the world *should* be, only how the world *may* be. Each individual, company, and nation must come to see that it can affect the outcome. But if the sustainable world is to come about, all companies, not just the best, will have to become lean and clean, as will a host of other organizations—the financial and investment community; international lending institutions; the scientific, engineering, and architectural communities; and local, state, and national governments. This is a future we must all invent.

INDUSTRIAL ECOLOGY

The future does not merely hold in store increased energy and resource efficiency for individual firms. Since firms cannot completely eliminate their waste streams, they will naturally seek productive uses for their waste and try to sell it for profit. Thus, the ultimate in multiple resource use and cascading goes well beyond a single company. Using nature as a model, we can attempt lean and clean design on a very large scale: the creation of an industrial ecosystem, in which individual companies act like individual organisms, and one company's waste becomes another's raw material. Two researchers at the General Motors Research Laboratories in Warren, Michigan, put it this way:

> The traditional model of industrial activity—in which individual manufacturing processes take in raw materials and generate products to be sold plus waste to be disposed of—should be transformed into a more integrated model: an industrial ecosystem. In such a system the consumption of energy and materials is optimized, waste generation is minimized and the effluents of one process—whether they are spent catalysts from petroleum refining, fly and bottom ash from electric-power generation or discarded plastic containers from consumer products—serve as the raw material for another process.[12]

One of the most detailed cases of industrial ecology is in the town of Kalundborg, Denmark. Cooperation has evolved among a dozen diverse enterprises, including an electric power plant, an oil refinery, district heating, a biotechnology production plant, a plasterboard factory, a sulfuric acid producer, cement producers, and local agriculture and horticulture operations (see the figure on page 174). The Kalundborg discussion is based on the work of Hardin Tibbs, a consultant with Global Business Network.[13]

In the early 1980s Asnaes—Denmark's largest coal-fired electrical power plant—started supplying process steam to the Novo Nordisk pharmaceutical plant and the Statoil refinery. Asnaes also started supplying surplus heat to support Kalundborg's district heating, eliminating the need for 3500 domestic oil-burning heating systems. Before this the company had been condensing the steam and releasing it into the local fjord. Water conservation is important in Kalundborg because freshwater is scarce and must be pumped from a lake several miles away. The refinery supplies cooling water and purified wastewater to Asnaes, which will soon also use purified wastewater from the pharmaceutical plant.

The refinery had been selling some of its surplus gas to Gyproc, the wallboard producer, since the early 1970s. In 1991 the refinery started selling its remaining surplus gas to Asnaes, saving 30,000 tons of coal a year. This was now possible because the refinery had begun removing the excess sulfur in the gas to make it cleaner burning. The leftover sulfur is sold to Kemira, which runs a sulfuric acid plant in Jutland.

Asnaes will also be removing sulfur from its smoke. The desulfurization process generates calcium sulfate, 80,000 tons of which will be sold each year to Gyproc as "industrial gypsum"—a substitute for the mined gypsum the company now imports. In addition, Asnaes uses its excess heat for warming its own seawater fish farm, which produces 200 tons of trout and turbot a year for the French market. Local farmers use sludge from the fish farm. The coal plant has even more surplus heat available and is considering using it for a thirty-seven-acre horticulture operation under glass. Asnaes's fly ash is used for cement making and road building. Local farms also use 330,000 tons a year of high nutrient-value sludge from the fermentation operations at the pharmaceutical plant as a liquid fertili-

THE FLOW OF RESOURCES AND BY-PRODUCTS BETWEEN PARTICIPANTS
IN THE INDUSTRIAL ECOSYSTEM AT KALUNDBORG

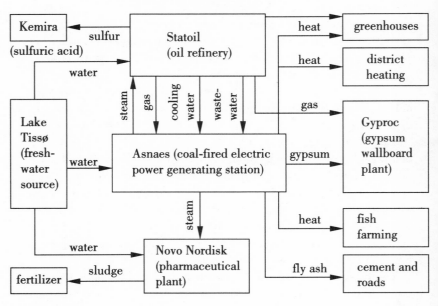

Source: Novo Nordisk.

zer. This type of sludge is "normally regarded as waste," but the plant treats it to neutralize any remaining microorganisms.

This astonishing cooperation was not required by regulation. Nevertheless, regulations such as those mandating reduced nitrogen in wastewater encouraged the cooperation that led to waste reuse. Each exchange was negotiated independently, and in some cases one company would install the needed infrastructure in return for another company's offer of a good price. Tibbs notes that Kalundborg could be a model for small and midsized cities around the world. Most such exchanges are between participants that are geographically close. Infrastructure costs are often a major factor, as in the case of heat transfer.

Tibbs also notes that the industrial ecosystem "focuses on the efficient interchange of byproducts and intermediates between industrial players, which roughly correspond to the individuals of a species in the biological ecosystem." The parallel concept "industrial metabolism" is "concerned with the efficiency of the metabolic processes occurring within the species' individuals, which roughly correspond to individual firms or industrial process operations."[14] Lean and clean production—pollution prevention and resource efficiency—is aimed at improving industrial metabolism.

Industrial ecology will develop naturally, if slowly, as more and more companies improve their industrial metabolism. A key group may be salespeople, who will require training to identify markets for waste and will need to be part of process redesign teams to help ensure that design decisions lead a company to waste streams that another company wants to buy. Public- and private-sector researchers can help by making industrial ecology considerations an explicit component of their efforts, which will ensure that clean companies shift not merely toward more benign outputs but also toward outputs that can be used as inputs by another company.

The industrial ecosystem represents the ultimate stage in the next industrial revolution, the switch from economies of scale to ecologies of scale.

Conclusion

MANAGEMENT, THE ENVIRONMENT, AND JOBS

There is a widespread sense that Western economies are moving into a post-industrial phase in which many lost jobs aren't coming back—and that current political leaders, geared to a Cold War military-industrial complex, don't know how to generate new ones.
—*The Wall Street Journal*, July 3, 1992[1]

Middle-class workers are under assault from global economic forces that seem beyond the reach of virtually any government policy. We now know that every wealthy country in the world is having trouble creating jobs.
—BILL CLINTON, speech to AFL-CIO, October 1993

Pollution prevention may be the solution to America's jobs problem. Lean and clean may also be the solution to the nation's slow productivity growth, which has kept wages low. Before considering the benefits to the nation as a whole, let's review a few cases in which individual companies used lean and clean techniques to save jobs or cut costs significantly without firing people or reducing wages:

- Dow Chemical trains its Louisiana workers in continuous process improvement to prevent waste. Since 1982 those workers have put in place hundreds of projects that save the company more than $110 million a year in nonlabor costs. The average return on investment of those projects was 204 percent—a six-month payback.

- A Sealtest ice cream plant on the verge of being shut down put in place a utility-financed modernization that reduced energy costs from 7.5 to 5.5 cents per gallon of ice cream and increased productivity 10 percent. That competitive edge saved 200 jobs and the plant is now adding new manufacturing workers.

• A comprehensive efficiency program saved the Southwire Corporation $40 million in energy costs from 1981 to 1988. During a rough financial period, the savings were almost equal to all the company's profits and may well have saved the company: a total of 4000 jobs at ten plants in six states.

• Republic Engineered Steels kept layoffs to a bare minimum during the recent slowdown in steel by putting in place an employee suggestion system. The resulting suggestions, many aimed at recycling and pollution prevention, have already reduced annual costs $45 million and will eventually cut another $20 million or more. The company has now tied wage increases directly to the employees' ability to reduce nonlabor costs.

Widespread adoption of lean and clean management would have a tremendous effect on our nation's troublesome job outlook in this so-called postindustrial era: Lean and clean techniques would save hundreds of thousands of jobs that will otherwise be lost to foreign competition. Replacing resource use with human labor and brainpower—and increased manufacture of energy- and resource-efficient products—would create hundreds of thousands of new jobs. The productivity gains would keep wages high.

Macroeconomically, energy efficiency and pollution prevention generate jobs for several reasons. First, efficiency and prevention lead companies to take some of the money they had devoted to consuming energy and resources and shift it to investing in technology and capital equipment. This shift creates economic growth and jobs. Indeed, shifting from consumption to investment may be the single most important transformation the U.S. economy must undergo if we are to remain prosperous in the next century.

Second, when a company cost-effectively reduces its costs for natural resources, waste disposal, and cleanup, it can lower the price of its products. This increase in competitiveness allows the company to beat foreign competitors both in the U.S. market and abroad, decreasing imports and increasing exports, creating even more jobs. As we have seen, pollution prevention is very cost-effective, with an annual return on investment that can exceed 50 percent to 100 percent. Competitiveness is further enhanced by the increase in productivity that invariably accompanies pollution prevention. (And U.S. businesses and consumers who buy the product of a clean producer at a lower cost will also have more money to spend on capital investment and savings.)

Third, as discussed in the previous chapter, the world market for energy-efficient and pollution prevention technologies is projected to be one of the fastest growing markets in the 1990s and beyond. A rich export market awaits the nation that leads in developing clean technologies.

Fourth, preventing pollution reduces all the negative economic consequences of environmental degradation: the loss of valuable natural resources such as clean water, adverse health effects, destruction of property (through corrosion, for instance), damage to the ecosystem, and reduced quality of life. Although difficult to quantify, these costs to the nation are of paramount importance to national prosperity. After all, the ultimate goal for the nation is not economic growth per se but economic growth that translates into sustainable increases in living standards and quality of life.

And, fifth, a great deal of capital now used very inefficiently by the economy will be freed up for more productive purposes by pollution prevention. The negative consequences of pollution cause an enormous diversion of national wealth. Consider all the capital that goes toward end-of-pipe pollution controls, waste disposal, and cleanup—capital that does not increase economic productivity and often has only a limited impact on the environment. Pollution prevention avoids the environmental damage, increases economic productivity, and frees up capital.

A 1993 analysis for the U.S. Department of Energy attempted to quantify the macroeconomic benefits of pollution prevention.[2] The results:

A 10%–20% reduction in waste by American industry would generate a cumulative GDP increase between 1996 and 2010 of $1.94 trillion. By 2010, the improvements would be generating 2,000,000 net new jobs—equal to roughly 1.5% of employment in that year. The study noted that this is "a relatively large impact considering that the investments driving it were assumed to be made for purposes other than increasing employment."

This analysis is conservative, since the study attempted to quantify only the first two previously listed benefits of reducing pollution. Imagine how much the economy would grow, how our competitiveness would increase, how exports would expand, how many jobs would be created, and how much better our environment would be if the nation reduced industrial waste 50 percent or more.

Pollution prevention represents a source of enduring competitive advantage for an industrial nation. Waste prevention skills can be upgraded continuously, and processes can constantly be improved: The savings never end. In Dow Chemical's case, even after ten years and nearly 700 projects, the 1992 and 1993 waste reduction contests had 249 winners with an average return of 300 percent. Pollution prevention skills are high-wage, knowledge- and process-based skills that are difficult for developing nations to duplicate. Such skills offer the possibility for long-term competitive success in the global economy, as made clear in many books, such as *The Work of Nations* by Robert Reich.

Clearly, pollution prevention should be one of the highest goals of public policy. Few public investments can help the country increase living standards more than cooperative research with industry to develop clean technologies and industrial processes that reduce the nonlabor costs of doing business. Federal laboratories seeking a post–Cold War mission could find no more important one than promoting economic, energy, and environmental security through the development of environmentally conscious processes, technologies, and materials. Local, state, and federal governments also need to disseminate information to businesses about the availability of cost-effective clean technologies and financing. More companies need to learn about utility rebate programs and about the Small Business Administration's multibillion-dollar loan program, some of which can be used to finance energy efficiency or pollution prevention improvements.

One fundamental reason why lean and clean techniques are not widely used is that they require cross-disciplinary teams, and our university system is too focused on narrow training in individual disciplines. As but one example, architects need more training in engineering, engineers need more training in architecture, and both need team training in energy-efficient design. Graduate schools, apprenticeship programs, and government training and retraining programs should include training in energy efficiency, waste prevention, and team-based process improvement.

Finally, the tax code could be improved to promote the purchase of capital equipment used for energy efficiency and pollution prevention. The Dutch allow firms that install innovative pollution prevention technologies to depreciate their investments in one year instead of ten. The tax break is available only for a list of technologies revised annually by a group of industry and government experts; technologies are removed from the list when they are required by regulation or when they gain a significant market share.[3] Promoting any cost-effective technology through the tax code generates the economic benefits associated with increased capital expenditures. Unlike some technologies, however, clean technologies do not replace labor, hence they do not decrease income tax revenues. Moreover, energy and resource bills are expensed by companies on their tax returns; clean technologies reduce energy and resource bills, thereby increasing tax revenues. Accelerated depreciation of clean technologies (or a clean technology investment tax credit) would generate more revenues than any other technology-oriented change in the tax code. *A well-designed tax change would probably be a net revenue raiser.*

While public policy and public investment are necessary to increase the potential for pollution prevention, that potential for nationwide gains in jobs and reductions in resource use and pollution will always exceed the reality. The most important obstacle to lean and clean lies in management, not government: It is unlikely that even the majority of companies will ever adopt the practices of the best companies described in this book.

Any company, however, can become far more profitable and productive by embracing lean and clean management. The change required to do so will not be easy because it must be systematic. It must involve all production and service delivery processes, which means it must involve all managers and employees and all their relationships with one another, with customers, and with suppliers.

The "secret" to success—to achieving a significant and enduring increase in profits and productivity—is product and process redesign. The focus of the redesign effort should be minimizing or eliminating waste. Japanese companies, and the best American ones, have achieved their tremendous successes by reducing wasted time, by preventing defects. Reducing wasted resources—preventing pollution—leads to Japanese-style process redesign, which is why it is invariably accompanied by a dramatic gain in productivity.

Combining the two approaches—eliminating wasted time *and* wasted resources—is the secret to lean and clean management. It was the secret to Henry Ford's success. And it can be the secret to your success, too.

YOU JUST DON'T UNDERSTAND: U.S. MISPERCEPTIONS OF JAPANESE SUCCESS

A *Wall Street Journal* front-page article from May 1993 makes clear just how completely U.S. companies have misunderstood what the Japanese have done:

> The current effort to redesign areas of the manufacturing landscape reflects a stunning admission of failure. Nearly 10 years after rushing to copy the Japanese, many American companies have failed to duplicate, let alone surpass, their efficiency. The American companies were hampered by cultural differences and, they acknowledge, by misapplication of what they saw in Japanese plants.[1]

Top American companies thought the Japanese succeeded merely by automating and reducing inventories rather than systematically redesigning their processes. A systemically flawed company that automates merely becomes an automated flawed company, and a flawed company that reduces inventory becomes a flawed company with less inventory. Worse still, automation and inventory reduction may well exacerbate the company's systemic problems. Chapter 1 explained how U.S. companies ended up focusing on operations, while Japanese companies focused on process.

For example, Federal-Mogul "mistakenly surmised that the Japanese got a major cost advantage from computers, robots and other automated equipment." In 1987 the company revamped one of its auto-parts plants with state-of-the-art automation. Soon, however, the company found that while it could produce parts much faster than before, it couldn't change product lines quickly. Switching from producing a small clutch bearing to a large one "required a slew of changes, ranging from readjusting parts 'feeding' systems to realigning the mechanisms that hold parts in place while they're being machined." Also, the complex machinery required extensive maintenance.

The result: Automation didn't lower costs and it actually reduced the company's ability to respond quickly to customer needs in an automotive market marked by increasing proliferation of new models. In the words of Fred Musone, the president of the chassis group, "We lacked flexibility." How ironic that Federal-Mogul became *less flexible* trying to imitate the Japanese, even though the Japanese process it was trying to copy is called flexible manufacturing.

Complexity and automation are not the answer. The answer is intelligent process redesign to make work easier for employees. Eventually Federal-Mogul revamped the plant using these principles. Now, for instance, "the parts that workers need are kept in bins within easy reach." And when Musone walks through the plant, he "shows off its 'simplicity.'" The new result: The plant produces three times as many varieties as before in the same amount of time. Because it can change quickly, it produces only what its customers need when they need it. This is end-use or demand-pull manufacturing.

Another example from the article concerns General Electric's experience with quality circles. The American company somehow found the "Japanese system too rigid and narrow for freewheeling American workers." In the original version of GE's quality circles, "workers discussed defined topics in dozens of different areas, often isolated from each other, and they get so much direction from the top that their contributions are seldom substantial." One GE division manager was quoted as saying, "Japanese companies are very hierarchical; we aren't. American workers don't stand up and salute." Ultimately, GE switched to a system in which "workers propose ideas that may require significant change and investment," and in some cases workers are given cash awards. In other words, GE apparently thinks it failed at doing it the Japanese way and thinks it succeeded at doing it the U.S. way—when exactly the reverse is true. The company ended up with a suggestion system that sounds suspiciously similar to what the best Japanese companies do (see Chapter 3).

Finally, the article noted, many U.S. companies were confused about just-in-time and inventory reduction. In the words of John Cassidy, director of research at United Technologies, many U.S. companies are learning that "the primary motive in just-in-time isn't reducing inventory." Cassidy, who spent a year studying manufacturing in Japan, notes that the Japanese devised just-in-time "to expose the weak points on the manufacturing line," to remove the inventory buffer and thereby force workers to keep production defect free. According to Cassidy, "We misplaced the goal. In order to eliminate inventory, we focused our efforts on materials-handling. We force the supplier to take extraordinary measures to solve our inventory problem, rather than looking at our manufacturing process."

These American companies focused on operations rather than processes. They viewed Japan's success at process redesign through the flawed lens of operations redesign and traditional industrial engineering. Success requires avoiding this classic mistake, which in turn requires understanding what the Japanese were *really* doing and what U.S. companies need to do to duplicate and surpass them.

Notes

Preface: The Fastest in the West—and the Greenest

1. The case study presented in this chapter is based on interviews of Robert McLean and Lee Windheim and Postal Service data developed by McLean.

2. "Inside Post Offices: The Mail Is Only Part of the Pressure," *New York Times*, May 17, 1993, p. A1.

Introduction: Lean and Clean Management

1. Michael Porter, Keynote Address before EPA's Clean Air Marketplace Conference, September 8, 1993.

2. Carol L. Fischer and Walter H. Zachritz, *A Review of Industrial Waste Minimization Case Studies*, Southwest Technology Development Institute, Waste-Education Research Consortium, New Mexico State University, Las Cruces, December 1992. Fischer and Zachritz examined numerous data bases on waste reductions, containing hundreds of actual case studies covering a wide range of industry sectors and project sizes. From that group seventy-five cases were chosen that had comparable detail on economic performance, such as payback. The Dow cases come from Kenneth E. Nelson, "Dow's Energy/WRAP Contest," a paper prepared for the 1993 Industrial Energy Technology Conference, March 24 and 25, 1993, Houston, TX. Nelson is a former energy conservation manager for Dow U.S.A.

For the purposes of this book, return on investment (ROI) = annual savings or earnings multiplied by 100 and divided by project costs.

3. Cited in Joel Makower, *The E-Factor* (New York: Times Books, 1993), p. 66.

4. Michael Porter, "America's Green Strategy," *Scientific American,* April 1991, p. 168.

5. Labor Secretary Robert Reich cited two such studies in "Companies Are Cutting Their Hearts Out," *New York Times Magazine,* December 19, 1993, pp. 54–55. Reich wrote, "Although they do not register directly on the balance sheets, high employee morale and loyalty are often among the most important of a firm's assets. Paring the payroll may imperil these assets in ways that escape profit and loss statements but profoundly affect competitive advantage."

6. U.S. Congress, Office of Technology Assessment, *Building Energy Efficiency,* OTA-E-518 (Washington, DC: U.S. Government Printing Office, May 1992), pp. 84–85, citing P. Komor and R. Katzev, "Behavioral Determinants of Energy Use in Small Commercial Buildings: Implications for Energy Efficiency," *Energy Systems and Policy,* vol. 12, 1988, p. 237.

7. *International Quality Study,* American Quality Foundation and Ernst & Young, 1991.

8. Gavin Wright, "The Origins of American Industrial Success, 1879–1940," *American Economic Review,* September 1990, p. 651.

9. "Some Companies Cut Pollution by Altering Production Methods," *Wall Street Journal,* December 24, 1990, pp. 1, 21.

10. U.S. Congress, Office of Technology Assessment, *Industry, Technology, and the Environment: Competitive Challenges and Business Opportunities,* OTA-ITE-58 (Washington, DC: U.S. Government Printing Office, January 1994), pp. 244–50. Those who believe that most companies have already taken advantage of most profitable pollution prevention opportunities should read the section "Factors Limiting the Adoption of Pollution Prevention," as well as the entire report.

11. Thomas Kuhn, *The Structure of Scientific Revolutions,* 2d edition (Chicago: University of Chicago Press, 1970), p. 77.

12. Deming quotation from the back jacket of Peter Senge, *The Fifth Discipline* (New York: Doubleday, 1990), p. 73.

Chapter 1: The American Origins of Lean and Clean

1. Henry Ford, *Today and Tomorrow* (Reprint, Cambridge, MA: Productivity Press, 1988), pp. 112, 113. The book was originally published by Doubleday, 1926.

2. Ecclesiastes 1:9–11.

3. David Halberstam, *The Reckoning* (New York: Avon, 1986). This book is an especially readable and detailed history of the U.S. auto indus-

try, especially Ford, paralleled with the history of the Japanese auto industry, particularly Nissan.

4. Ford, *Today and Tomorrow,* p. 126. Ford noted that "the tradition of lumbering is of waste—that is why wages are so low and the prices of timber so high."

5. Ibid., p. 136.

6. Ibid., pp. 92–100, for the steel and blast furnace examples.

7. The discussion of interchangeable parts and the moving assembly line, as well as the cycle time numbers, comes from James P. Womack et al., *The Machine That Changed the World* (New York: Rawson Associates, 1990).

8. Ford's discussion of time can be found in his *Today and Tomorrow,* pp. 102–21.

9. Shigeo Shingo, *Study of the Toyota Production System* (Tokyo: Japanese Management Association, 1981), pp. 117–28, as cited in Robert Hayes et al., *Dynamic Manufacturing* (New York: Free Press, 1988), p. 45.

10. Ford, *Today and Tomorrow,* p. 16. Spender quoted in Halberstam, *The Reckoning,* p. 80.

11. Ford, *Today and Tomorrow,* p. 6.

12. Ibid., p. 52.

13. Halberstam, *The Reckoning,* p. 84.

14. Ibid., p. 81.

15. David Garvin, *Managing Quality* (New York: Free Press, 1988), pp. 182–83. Chapter 10, "The Japanese Quality Movement," is an excellent discussion of some of the factors in Japan that helped them to embrace quality.

16. *Car Talk,* National Public Radio, April 11, 1993.

17. Shigeo Shingo, *Modern Approaches to Manufacturing Improvement: The Shingo System,* ed. Alan Robinson (Cambridge, MA: Productivity Press, 1990), pp. 43–44. This book is an outstanding collection and integration of Shingo's writing. Shingo wrote that his teacher drilled into him the ideas of Taylor and especially the Gilbreths, and that they "lie at the heart of my improvement activities."

18. Ibid., pp. 35, 44.

19. Hayes et al., *Dynamic Manufacturing,* p. 58.

20. Shingo's discussion of the Industrial Revolution comes from Shigeo Shingo, *The Shingo Production Management System* (Cambridge, MA: Productivity Press, 1992), pp. 7–8.

21. Ibid., p. 54.

22. Shingo (1990), op. cit., p. 23.

23. Shingo (1992), p. 55, which discusses Gilbreth's error.

24. The examples are cited in Shingo (1990), op. cit., p. 9.

25. Robert Hayes and Kim Clark, *Interfaces,* 15 (1985), p. 13, as cited in Shingo (1990), p. 9.

26. Shingo (1992), p. 55.

27. Taiichi Ohno, *Toyota Production System,* translated (Cambridge, MA: Productivity Press, 1988), 97–109. The book was originally published in Japan by Diamond Inc., 1978. The remaining quotations in this section can be found on p. 59 (implementing Toyota system); pp. 19–20 (formula for success and types of waste); pp. 54–55 (inventory waste); and p. 17 (repeat *why* five times).

28. The numbers for the Toyota suggestion system come from Tom Peters, *Thriving on Chaos* (New York: Harper & Row, 1988), p. 88.

CHAPTER 2: SYSTEMS: THE CYCLE OF LIFE

1. Robert Reich, *The Work of Nations* (New York: Alfred A. Knopf, 1991), pp. 208, 213.

2. The discussion of Compaq is based on personal communications with Ron Perkins, as well as a case study: Ron Perkins and Ted Flanigan, "The Compaq Experience: Corporate Dynamics and Energy Efficiency" (Basalt, CO: IRT Environment, December 1992).

3. Donella H. Meadows, "Whole Earth Models and Systems," *Coevolution Quarterly,* Summer 1982, pp. 98–108.

4. Peter Senge, *The Fifth Discipline* (New York: Doubleday, 1990), p. 73.

5. Ibid., pp. 71–72.

6. Ibid., p. 72.

7. This story comes from Todd Campbell, "From Megawatts to Nega-Watts," *Horizon Air Magazine,* November 1991, pp. 17–37, as well as personal communications with Ron Gonsalves.

8. These principles distill Edward Deming's fourteen points for Total Quality Management and Stephen Covey's "Seven Habits of Highly Effective People." Their other principles can be derived from these.

9. "Chemical Firms Find That It Pays to Reduce Pollution at Source," *Wall Street Journal,* June 11, 1991, pp. A1, A6.

10. The quotation from Artzt comes from Bruce Smart, ed., *Beyond Compliance: A New Industry View of the Environment* (Washington, DC: World Resources Institute, 1992), pp. 35–37.

11. As cited in Smart, *Beyond Compliance*, p. 35.

12. Amory Lovins, *Soft Energy Paths* (New York: Harper & Row, 1979), p. 39.

13. Masaaki Imai, *Kaizen* (New York: McGraw-Hill, 1986), p. xxix.

14. MIT Commission on Industrial Productivity, *Made in America: Regaining the Productive Edge* (Cambridge, MA: MIT Press, 1989), p. 49.

15. For a discussion of reengineering and its relationship to lean and clean management, see Richard Wells et al., "What's the Difference Between Reengineering and TQEM?" *Total Quality Environmental Management*, Spring 1993, pp. 273–82.

16. The Dow story comes from Kenneth E. Nelson, "Dow's Energy/WRAP Contest," a paper prepared for the 1993 Industrial Energy Technology Conference, March 24 and 25, 1993, Houston, TX.

17. Donella Meadows, The Global Citizen, "The Importance of Accurate Feedback," *Valley News*, August 17, 1991.

18. "Some Companies Cut Pollution by Altering Production Methods," *Wall Street Journal*, December 24, 1990, pp. 1, 21, and "Why AT&T Is Dialing 1 800 Go Green," *Business Week* (special issue on quality), October 25, 1991, p. 49.

19. Of course, as will be discussed later, if one's orientation is severely distorted, then even this kind of feedback can be misinterpreted.

20. The phrase *feedback of experience* is from Philip R. Thomas, *Competitiveness Through Total Cycle Time* (New York: McGraw-Hill, 1990), p. 9.

21. "Concurrent Engineering," *Aerospace America*, March, 1993.

22. Mark Dorfman et al., *Environmental Dividends: Cutting More Chemical Wastes*, a report published by INFORM in New York, 1992, pp. 172–81.

23. The results of the study are given in two reports by the Secretary's Commission on Achieving Necessary Skills: *What Work Requires of Schools* and *Skills and Tasks for Jobs* (Washington, DC: U.S. Department of Labor, 1991).

Chapter 3: Labor and Lean and Clean

1. Cited in Japan Human Relations Association, *The Idea Book* (Cambridge, MA: Productivity Press, 1988), p. 44.

2. The discussion of Republic Engineered Steels is based on personal communication with Harold Kelly and the following articles: "Stanching the Loss of Good Jobs," *New York Times*, January 31, 1993, pp. F1, F6;

"Republic Charts New Waters with a Tightly Run Ship," *Metal Bulletin Monthly*, November 1991; "New Paths in Business When Workers Own," *New York Times*, November 22, 1991; "Republic Attempts Recasting," *Beacon Journal*, August 26, 1991, p. C1; "A Democratic Republic," *Incentive*, February 1992; and "Manufacturing Excellence Awards," *Controls and Systems*, January 1992.

3. Henry Ford, *Today and Tomorrow* (Reprint, Cambridge, MA: Productivity Press, 1988), p. 124.

4. Japan Human Relations Association, *Idea Book*, pp. 201–2.

5. Shingo (1990), op. cit., p. 13.

6. Masaaki Imai, *Kaizen* (New York: McGraw-Hill, 1986), pp. 111–24.

7. Japan Human Relations Association, *The Idea Book*, p. 190.

8. Imai's book, *Kaizen*, is also useful.

9. "Maier Leads Republic Efforts to Survive," *Repository*, Canton, OH, November 25, 1991.

10. Japan Human Relations Association, *Idea Book*, p. 120.

11. A basic introduction to quality control tools can be found in ibid., pp. 21–30.

12. Beverly Geber, "Saturn's Grand Experiment," *Training*, June 1992, p. 32.

13. David Garvin, *Managing Quality* (New York: Free Press, 1988), pp. 202–3.

14. Japan Human Relations Association, *Idea Book*, p. 80.

15. Alan Lightman and Owen Gingerich, "When Do Anomalies Begin?" *Science*, February 7, 1991, pp. 690–95.

16. The Sealtest case is based on conversations with Joseph Crowley, the plant's manager, and John Donoghue, manager of the load containment division of Boston Edison; a paper written by Crowley and Donoghue, "The Energy Efficiency Partnership: Kraft General Foods and Boston Edison Company," presented to the American Council for an Energy-Efficient Economy Workshop on Partnerships for Industrial Productivity through Energy Efficiency, September 20, 1993; data provided by Donoghue; Ross Gelbspan, "At Sealtest, Sweet Smell of Success with Energy," *Boston Globe*, October 9, 1991, p. 37; and "The Energy Efficiency Industry and the Massachusetts Economy," a report from the Massachusetts Energy Efficiency Council, Concord, MA, December 1992 (which bases some of its discussion on a Kraft General Foods presentation to Secretary James Watkins and Governor William Weld in Framingham, MA, on February 13, 1992).

CHAPTER 4: A FOCUS ON PREVENTION: THE CASE OF COMPAQ COMPUTER

1. The Compaq study presented here comes primarily from personal communications with Ron Perkins; Ron Perkins and Ted Flanigan, "The Compaq Experience: Corporate Dynamics and Energy Efficiency" (Basalt, CO: IRT Environment, December 1992); and a discussion of Compaq in "The New American Century," a special issue of *Fortune*, Spring–Summer 1991, pp. 27–28.

CHAPTER 5: ENERGY EFFICIENCY

1. Amory Lovins, *Soft Energy Paths* (New York: Harper & Row, 1979), p. 39.

2. U.S. Congress, Office of Technology Assessment, *Building Energy Efficiency*, OTA-E-518 (Washington, DC: U.S. Government Printing Office, May 1992), pp. 84–85, citing P. Komor and R. Katzev, "Behavioral Determinants of Energy Use in Small Commercial Buildings: Implications for Energy Efficiency," *Energy Systems and Policy*, vol. 12, 1988, p. 237.

3. *Office Access* (San Francisco: Harper Perennial, 1992), p. 53.

4. *Green Lights* (Washington, DC: Environmental Protection Agency, July 1993).

5. The discussion of Boeing is based on a visit to some of the company's Washington State buildings, personal conversations with Lawrence Friedman and Steve Cassens, articles in *Boeing News* from May 10, 1991, and January 15, 1993, and 1992 EPA data on the "Green Lights" program. It should be noted that the local utility covers about 75 percent of the cost of Boeing's new lighting with rebates, which has sped up the payback. On the other hand, Boeing calculates the payback on the basis of the low-cost electricity available in the Pacific Northwest, about 3.5 cents per kilowatt-hour, which is almost half the national average. Some companies pay three times what Boeing does for electricity.

6. U.S. Congress, Office of Technology Assessment, *Energy Efficiency in the Federal Government: Government by Good Example?* OTA-E-492 (Washington, DC: U.S. Government Printing Office, May 1991), p. 31. Many utilities promote efficiency because of the tremendous potential savings involved: The electricity savings in lighting alone would reduce the need for about 70–120 billion watts of power plants that would cost $85–$200 billion to build and $18–$30 billion per year to operate.

7. "Annual Report on American Industry," *Forbes*, January 4, 1993, p. 134.

8. Arnold Fickett, Clark Gellings, and Amory Lovins, "Efficient Use of

Electricity," *Scientific American*, September 1990. The article is a consensus document between Rocky Mountain Institute (Amory Lovins is the research director) and the utility think tank the Electric Power Research Institute (Arnold Fickett and Clark Gellings are, respectively, vice president and director of EPRI's customer service division).

9. Jim Rogers, "Energy Management Case Studies," a paper presented at the Competitek Forum, September 1991, Snowmass Village, CO.

10. There is some labor cost in removing the unneeded overhead lamps, but that is probably more than covered by the reduced maintenance required by fewer lamps.

11. Paul Scanlon, "Effective Management Cuts Costs and Helps Environment," *AS&U*, September 1991, pp. 40–42.

12. The Hyde Tools story is based on a report in the *TPM Newsletter*, January 1993, p. 7; personal communications with Doug DeVries; and Joseph E. Paluzzi, and Timothy J. Greiner, "Finding Green in Clean: Progressive Pollution Prevention at Hyde Tools," *Total Quality Environmental Management*, Spring 1993, pp. 283–90.

13. This case study is from the National Lighting Bureau, *Office Lighting and Productivity* (2101 L Street, NW, Suite 300, Washington, DC 20037), 1988, as well as the bureau's press release on the lighting upgrade, January 31, 1985.

14. This section is based on Amory Lovins and Rick Heede, *The State of th Art: Appliances*, Competitek report, 1990. See also Joel Makower, *The E-Factor* (New York: Times Books, 1993), pp. 225–29.

15. New Hewlett Packard printer described in *New York Times*, May 16, 1993, p. F12. Fred Forsyth quoted in *Quad Report*, April 1993, p. 2.

16. "Cost Reduction Through Conservation," *Inc.*, March 1992, pp. 109–10.

17. Ken Teeters's remarks to the E-Source Forum, September 30, 1992, Snowmass Village, CO.

18. The Comstock case study is based on U.S. Congress, *Building Energy Efficiency*, p. 51; personal communications with Paul Scanlon; and P. Rittelmann and P. Scanlon, "HVAC Design Delivers Twin Benefits," *Building Design and Construction*, November 1984.

19. Amory Lovins, "Energy-Efficient Buildings: Institutional Barriers and Opportunities," a strategic issue paper published by E Source, Boulder, CO, 1992, p. 11. This is probably the most comprehensive analysis of why so few energy-efficient buildings get built.

20. Ibid. and personal communications with Scanlon.

Chapter 6: Lean and Clean Design for Office Productivity

1. Frank Lloyd Wright is quoted in Michael D. Shanus et al., "Going Beyond the Perimeter with Daylight," *Lighting Design & Application*, March 1984, pp. 30–40.

2. This case is based on Allen Russell, "Pennsylvania Power and Light: A Lighting Case Study," *Buildings*, March 1982, pp. 49–56; "Office Lighting Retrofit Will Pay Back in 69 Days," *Facilities Design & Management*, June, 1982, p. 13.

3. The Wisconsin case study is from the National Lighting Bureau, *Office Lighting and Productivity* (2101 L Street, NW, Suite 300, Washington, DC 20037), 1988. The National Lighting Bureau has numerous publications detailing the energy and productivity savings from improved lighting.

4. See William J. Dickson and F. J. Roethlisberger, *Counseling in an Organization: A Sequel to the Hawthorne Researches* (Cambridge, MA: Harvard University Press, 1966). This book explains in detail that the traditional, simplistic view of the Hawthorne effect—that changing the workplace environment affects productivity only because it signals management interest in the worker—is not at all what the Hawthorne researchers concluded from their work. For a survey of some of the literature on the flaws in the Hawthorne effect research, see Michael Brill et al., *Using Office Design to Increase Productivity*, vol. 1 (Buffalo: Workplace Design and Productivity, 1984), pp. 224–25. Brill offers the study by the Buffalo Organization for Social and Technological Innovation, which is presented in his book, as data that contradict the results of the Hawthorne experiments. These two works together make clear that the work of the Hawthorne researchers remains exceedingly misunderstood and misinterpreted.

5. The life-cycle numbers are cited in Shanus et al., "Going Beyond the Perimeter," p. 39.

6. The NMB case comes from William Browning, "NMB Bank Headquarters," *Urban Land Institute*, June 1992, and personal communications with Browning.

7. The Lockheed case study is based on Charles C. Benton and Marc Fountain, "Successfully Daylighting a Large Commercial Building: A Case Study of Lockheed Building 157," *Progressive Architecture*, November 1990, pp. 119–21; "Employees Respond to Lockheed Building 157," *Professional Energy Manager*, July 1984, p. 5; "Lockheed's No. 157: Ex Post Facto," *Facilities Planning News*, October 1984; and personal communications with Lee Windheim and Don Aitken.

8. The Wal-Mart case study is based on the Rocky Mountain Institute's

design consulting for and analysis of the Eco-Mart, and personal communications between Bill Browning and Tom Seay, Wal-Mart's vice president for real estate.

9. The discussion of West Bend Mutual is based on Paul Beck, "Intelligent Design Passes IQ Test," *Consulting-Specifying Engineer*, January 1993, pp. 34–38; and Walter Kroner et al., "Using Advanced Office Technology to Increase Productivity" (Troy, NY: Center for Architectural Research, 1992).

10. The RPI researchers noted, "Subjects were not informed that an analysis of their productivity was being conducted by the research team," and "Since the company's productivity measurements were ongoing and were not specifically noted by the employees, we believe that workers' behavior was not affected by their participation in the study."

11. Roger S. Ulrich, "Biophilia, Biophobia, and Natural Landscapes," in Stephen R. Kellert and Edward O. Wilson, eds., *The Biophilia Hypothesis* (Washington, DC: Island Press, 1993), pp. 73–137.

Chapter 7: Becoming Lean and Clean: A Systems Approach

1. George Stalk and Thomas Hout, *Competing Against Time* (New York: Free Press, 1990), pp. 110–14. The American competitor is not named.

2. "CEOs: The Vision Thing," *Wall Street Journal*, January 28, 1993, p. A14.

3. Tachi Kiuchi, remarks to the Eco Tech conference, March 4, 1994, Leesburg, VA, as well as personal communications.

4. For a more thorough discussion of Boyd's theory and its application to the Persian Gulf War and manufacturing, see Joseph Romm, *The Once and Future Superpower: How to Restore America's Economic, Energy, and Environmental Security* (New York: William Morrow, 1992), pp. 15–36. See also Romm, "The Gospel According to Sun Tzu," *Forbes*, December 11, 1992.

5. Roger W. Schmenner, "The Merit of Making Things Fast," *Sloan Management Review*, vol. 30, no. 1, 1989.

6. Abbie Griffin, "Metrics for Measuring Product Development Cycle Time," *J. Prod. Innov. Manag.*, vol. 10, 1993, pp. 112–25.

7. "The Force Was with Them: Army's Jedi Knights Forged Gulf War Strategy," *Boston Globe*, March 17, 1991, A23.

8. Sun Tzu, *The Art of War*, trans. Samuel B. Griffith (London: Oxford University Press, 1971).

9. Robert Fritz, *The Path of Least Resistance* (New York: Fawcett-

Columbine, 1989), as cited in Peter Senge, *The Fifth Discipline* (New York: Doubleday, 1990), p. 155.

10. *International Quality Study*, American Quality Foundation and Ernst & Young, 1991.

11. "Motorola Illustrates How an Aged Giant Can Remain Vibrant," *Wall Street Journal*, December 9, 1992, pp. A1, A20.

12. Robert Hayes, Steven Wheelwright, and Kim Clark, *Dynamic Manufacturing* (New York: Free Press, 1988), p. 156.

13. Senge, *Fifth Discipline*, p. 174.

14. Ian Mitroff, *Break-away Thinking* (New York: John Wiley, 1988), as cited in Senge, *Fifth Discipline*, p. 176.

15. Ibid.

16. James P. Womack et al., *The Machine That Changed the World* (New York: Rawson Associates, 1990), p. 77.

17. Senge, *Fifth Discipline*, p. 181.

18. All Pierre Wack's quotations come from two articles, "Scenarios: Uncharted Waters Ahead," *Harvard Business Review*, September–October 1985, pp. 73–89, and "Scenarios: Shooting the Rapids," *Harvard Business Review*, November–December 1985, pp. 139–50. As for others who foresaw the oil crisis, Charley Maxwell, a thirty-five-year-old oil expert, had been warning of an "energy crisis" since 1970. In June 1973 he met with executives from GM, Ford, and Chrysler to tell them it was imminent. They could not hear.

19. Both Roberta Wohlstetter and Mitsuo Fuchida are quoted in Wack, "Scenarios: Shooting the Rapids," pp. 148–49.

20. "A Very Old General May Hit the Beach with the Marines," *Wall Street Journal*, January 9, 1991. Peter Schwartz, *The Art of the Long View* (New York: Doubleday, 1991). Paul Hawken, *The Ecology of Commerce: A Declaration of Sustainability* (New York: Harper Business, 1993); Donella Meadows et al., *Beyond the Limits* (Post Mills, VT: Chelsea Green, 1992).

21. "Motorola Illustrates."

22. Personal communications.

23. "Motorola Illustrates."

24. The Harrah's case is based on Ken Teeters's remarks to the E-Source Forum, September 30, 1992, Snowmass Village, CO, and personal communications with Teeters.

25. The discussion of Home Depot comes from Patricia Sellers, "Companies That Serve You Best," *Fortune*, May 31, 1993, pp. 74–88. Data on the strategy experience of 3000 companies compiled by the Strategic Plan-

ning Institute show that the total direct costs of those companies with better support service are 3 percent *lower* that those of companies with worse service. As Phillip Thompson, a vice president of the institute, notes, "Better service quality is free!" See Thompson, "Strategic Management of Service," *Design Management Journal*, Winter 1992, pp. 62–70.

26. "Revenge of the 'Jedi,' " *Army Times*, April 22, 1991, pp. 12–14.

27. Womack et al., *Machine That Changed the World*, p. 92.

28. "Back to School for Honda Workers," *New York Times*, March 29, 1993, pp. D1, D3.

29. "Companies That Train Best," *Fortune*, March 22, 1993, pp. 62–75.

30. Arie De Geus quoted in Senge, *Fifth Discipline*, p. 4.

31. A key reason so many of America's biggest companies have slashed their middle management payroll in the last few years is that so many had an excess of white-collar employees. In the late 1980s, 77.5 percent of GM's workforce was white collar and salaried, 22.5 percent hourly blue-collar workers; Mobil Oil was 61.5 percent white collar; General Electric 60.0 percent; Du Pont, 57.1 percent; Chrysler, 44.4 percent; Exxon, 43.0 percent; AT&T, 42.0 percent. The figures are from Richard Rosecrance, "Too Many Bosses, Too Few Workers," *New York Times*, July 15, 1990, p. F11.

32. John Boyd, "Patterns of Conflict," unpublished lecture notes, June 1987, p. 74.

33. Marco Iansiti, "Real-World R & D: Jumping the Product Generation Gap," *Harvard Business Review*, May–June 1993, pp. 138–47.

34. Mark Dorfman et al., *Environmental Dividends: Cutting More Chemical Wastes* (New York: INFORM, 1992), p. 173.

35. The Taurus's safety record (death rate per 10,000 registered vehicles) is superior to that of the Honda Accord, the Toyota Camry, the Nissan Sentra, and virtually every comparable American car.

36. Womack et al., *The Machine That Changed the World*, p. 200.

37. Pollution Prevention Partnership, "Progress Report," Denver, CO, February 1993.

38. Shingo (1992), p. 55. A cogent discussion of "designing for resilience" can be found in Amory B. Lovins and L. Hunter Lovins, *Brittle Power* (Andover, MA: Brick House Publishing, 1982), pp. 177–213.

39. Nature also achieves resilience through diversity. The remarkable variety of life on the planet has allowed the global ecosystem to survive endless shocks. For biologist Edward O. Wilson, biological diversity is "the property that makes resilience possible" in nature. Preserving the tre-

mendous diversity of biological life is necessary to sustain the resilience of the planet's ecosystem. In an interlocking web, one can cut many strands and the web stays together. Cut too many, however, and it collapses. The fact that it has not collapsed yet provides no clue as to whether we are a few strands away from collapse or still quite secure. All we know is that the more nodes and the more interconnections, the more secure and resilient the system is.

40. Stuart Sutherland, "Evolution Between the Ears," *New York Times Book Review*, March 7, 1993, p. 16. Sutherland is professor emeritus of psychology at the University of Sussex.

41. "Motorola Illustrates."

CHAPTER 8: LEAN AND CLEAN DESIGN FOR FACTORY PRODUCTIVITY

1. Ford, *Today and Tomorrow* (Reprint, Cambridge, MA: Productivity Press, 1988), p. 94, and "Xerox: Design for Environment," Harvard Business School Case Study, N9-794-022, January 7, 1994, p. 20.

2. Mark Dorfman, "Source Reduction," *Pollution Prevention Review*, Autumn 1992, pp. 403–14. See also Mark Dorfman et al., *Environmental Dividends: Cutting More Chemical Wastes* (New York: INFORM, 1992).

3. Robert Hayes, Steven Wheelwright, and Kim Clark, *Dynamic Manufacturing* (New York: Free Press, 1988), pp. 176–77.

4. What makes the productivity rise "even more surprising" to the authors is that in these factories, this waste material "is a direct substitute for new raw materials and is recycled within the factory." In other words, reducing the waste through process improvement does not actually directly reduce material costs significantly because that wasted material would have been reused.

5. This hierarchy is modified from one put forth by EPA's David Wann, *Biologic* (Boulder, CO: Johnson Books, 1990), p. 106.

6. Pollution Prevention Partnership, "Progress Report," Denver, CO, February 1993.

7. The Martin Marietta case is based primarily on Kevin J. Dykema and George R. Larsen, "Case Study: The Greening of Corporate Culture: Shifting the Environmental Paradigm at Martin Marietta Astronautics Group," *Pollution Prevention Review*, Spring 1993, pp. 197–211.

8. Pollution Prevention Partnership, "Progress Report."

9. "Martin Marietta Employee Environmental Survey," Martin Marietta, Denver, CO, October 1991.

10. U.S. Congress, Office of Technology Assessment, *Industry, Technol-*

ogy, and the Environment: Competitive Challenges and Business Opportunities, OTA-ITE-58 (Washington, DC: U.S. Government Printing Office, January 1994), p. 236, which cites Ann Rappaport, *Development and Transfer of Pollution Prevention Technology Within a Multinational Corporation,* dissertation, Department of Civil Engineering, Tufts University, May 1992.

11. Braden R. Allenby, "Design for Environment: A Tool Whose Time Has Come," *SSA Journal,* September 1991, pp. 5–9.

12. U.S. Congress Office of Technology Assessment, *Green Products by Design: Choices for a Cleaner Environment* (Washington, DC: U.S. Government Printing Office), October 1992.

13. The historical discussion of GE Plastics and the UKettle is based on three case studies written by Karen J. Freeze and Dorothy L. Barton for the Design Management Institute of Boston, MA: "GE Plastics: Selecting a Partner" (1991), "Polymer Solution: Tempest About a Teapot" (1992), and "Great British Kettles: The UKettle Project" (1993). These cases detail the trials and tribulations the designers went through in this early effort at fast-cycle design for the environment. The UKettle is now sold as the Automatic Kettle Model AK-10 by MAXIM, which notes in its instructions and warranty that the kettle "is designed to encourage material recycling. All engineering plastics in the kettle are recyclable. At the end of the product life, the kettle's components can be disassembled and sorted by recycling professionals."

14. The design principles are based on Charles Leinbach, "Designing the Three Rs into a Product," *Innovation,* Special 1992 issue, p. 25. The discussion is also based on personal communications with Michael S. Jaeb of Fitch Richardson Smith and material provided by that company.

15. The phrase *feedback of experience* is from Philip R. Thomas, *Competitiveness Through Total Cycle Time* (New York: McGraw-Hill, 1990), p. 9. Faster cycle time means "increased opportunities to learn from the feedback of experience, which I call Cycles of Learning. Conscientious use of such feedback will, in turn, accelerate results."

16. John Sedgwick, "The Complexity Problem," *Atlantic Monthly,* March 1993, pp. 96–104.

17. Hayes et al., *Dynamic Manufacturing,* p. 174.

18. Ton Borsboom, "The Environment's Influence on Design," *Design Management Journal,* Fall 1991, pp. 42–47. The article has an excellent discussion of design-for-disassembly techniques.

19. "Design for Manufacturability at Midwest Industries," Harvard

Business School case study, No. 9-690-066, rev. February 6, 1992, p. 6.

20. Steven Ashley, "Designing for the Environment," *Mechanical Engineering*, March 1993, pp. 52–55.

21. "Purchasing the Design of Services," *Design Management Journal*, Winter 1992, p. 55.

22. Surveys cited in Sedgwick, "Complexity Problem," pp. 96–104. Akio Tanii quoted in "Japan's Electronics Firms Stress Product Simplicity," *Harrisburg Patriot News*, January 1, 1993, p. B5.

23. Sedgwick, "Complexity Problem."

24. The quotation is from Robert T. Lund, "Remanufacturing," *Technology Review*, February–March 1984, pp. 19–29. The discussion of remanufacturing draws material from this article as well as Craig Waters, "On Second Thought," *Inc.*, August 1984, pp. 54–61; "A Growing Love Affair with the Scrap Heap," *Business Week*, April 29, 1985, pp. 60–61; H. C. Haynsworth and R. Tim Lyons, "Remanufacturing by Design, the Missing Link," *Production and Inventory Management Journal*, 2d quarter, 1987, pp. 24–28; and Ashley, "Designing for the Environment," p. 54. The Haynsworth and Lyons piece discusses specific design for remanufacture techniques and lists numerous references.

25. The discussion of Xerox is based on "Xerox: Design for the Environment," Harvard Business School Case Study, N9-794-022, January 7, 1994.

26. E. Thomas Morehouse, Jr., "Design for Maintainability," American Electronics Association Design for the Environment White Paper, 1992.

27. David Cohan et al., "Beyond Waste Minimization: Life-Cycle Cost Management for Chemicals and Materials," *Pollution Prevention Review*, Summer 1992, pp. 259–75. See also Paul Bailey, "Life-Cycle Costing and Pollution Prevention," *Pollution Prevention Review*, Winter 1990–91, pp. 27–39.

28. Pollution Prevention Partnership, "Progress Report," p. 9.

29. The energy numbers throughout this section come from Steven Nadel et al., *Energy-Efficient Motor Systems* (Washington, DC: American Council for an Energy-Efficient Economy, 1991), and Arnold Fickett, Clark Gellings, and Amory Lovins, "Efficient Use of Electricity," *Scientific American*, September 1990.

30. Literally dozens of examples of dramatic energy savings with payback from one to three years (without counting utility rebates) were presented at a national conference on energy-efficient motor systems in Baltimore on February 9–10, 1993. The two cases in the text come from

"Energy Management Case Studies," a September 26, 1991, paper presented by Jim Rogers of EUA Cogenex Corporation (an energy services company) at the September 1991 Competitek Forum in Snowmass Village, CO.

31. Nadel et al., *Energy-Efficient Motor Systems*, pp. 90–93.

32. Taiichi Ohno, *Toyota Production System* (Cambridge, MA: Productivity Press, 1988), pp. 63, 126.

33. Peter Jaret, "Putting the Pinch on Energy Costs," *EPRI Journal*, July–August 1991, pp. 24–31.

34. The Southwire discussion is based on William D. Browning and L. Hunter Lovins, *The Energy Casebook* (Snowmass, CO: Rocky Mountain Institute, 1989), pp. 18–19; and on Nadel et al., *Energy-Efficient Motor Systems*, pp. 10, 80, 186, and 247.

35. Kenneth F. Dunker and Basile G. Rabbat, "Why America's Bridges Are Crumbling," *Scientific American*, March 1993, pp. 66–72.

36. Masaaki Imai, *Kaizen* (New York: McGraw-Hill, 1986), pp. 158–61.

37. Stephen Macaulay, "Amazing Things Can Happen If You . . . 'Keep It Clean,' " *Production*, May 1988, pp. 72–74.

38. Charles Garfield, *Peak Performers: The New Heroes of American Business* (New York: William Morrow, 1986), pp. 128–29, as cited in Herbert R. Steinbacher and Norma L. Steinbacher, *TPM for America* (Cambridge, MA: Productivity Press, 1993), p. 31.

39. See Macaulay, "Amazing Things Can Happen"; and Tom Harvey, "CMMS (Computerized Maintenance Management System) at Chrysler Assembly Plant," *Maintenance Technology*, March 1993, pp. 47–48.

40. Ron Moore et al., "Blueprint for Reliability," *Maintenance Technology*, March 1993, pp. 23–27.

41. Imai, *Kaizen*, pp. 158–161.

42. These indicators are derived from Moore et al., "Blueprint for Reliability," p. 26.

43. *TPM Newsletter*, January 1993, pp. 7–8.

44. The Haworth case comes from Larry Martin, *Proven Profits from Pollution Prevention, Vol. II* (Washington, DC: Institute for Local Self-Reliance, 1989), pp. 21–24.

45. The Baxter case comes from ibid., pp. 52–55, as well as personal communications with John Carter and information provided by Baxter.

46. Martin, *Proven Profits*, pp. 25–30.

47. Ford, *Today and Tomorrow*, pp. 132–33.

48. "An Engineer's Guide to Cogeneration," *Heating, Piping, Air Conditioning*, vol. 62, July 1990, pp. 83–98.

49. The discussion of cascading paper is based on GTS newsletters as well as "Cost Reduction Through Conservation," *Inc.*, March 1992, pp. 109–10, and "Ready to Recycle? Take These Five Steps to Success," *Modern Office Technology*, May 1992, p. 25.

50. Global Environmental Management Initiative (GEMI), *Environmental Self-assessment Program* (Washington, DC: GEMI, April 1993).

CHAPTER 9: THE FUTURE IS LEAN AND CLEAN

1. "The Commerzbank Report on German Business and Finance," advertisement in *Wall Street Journal*, October 22, 1993. Alan Kay is quoted in Curtis A. Moore and Alan S. Miller, "Green Gold," Center for Global Change, University of Maryland, College Park, September 1993 (a report based on the book of the same name).

2. *Environmental Technology Export Strategy*, Interagency Environmental Technologies Working Group, Washington, DC, 1993.

3. Joseph J. Romm, *The Once and Future Superpower: How to Restore America's Economic, Energy, and Environmental Security* (New York: William Morrow, 1992); Curtis A. Moore and Alan S. Miller, *Green Gold: Environmental Technology and the Race to Capture Industrial Dominance of the Twenty-first Century* (Boston: Beacon Press, 1994); and U.S. Congress, Office of Technology Assessment, *Industry, Technology, and the Environment: Competitive Challenges and Business Opportunities*, OTA-ITE-58 (Washington, DC: U.S. Government Printing Office, January 1994).

4. Carter Brandon and Ramesh Ramankutty, "Toward an Environmental Strategy for Asia: A Summary of a World Bank Discussion Paper" (Washington, DC: World Bank, 1993), p. 6.

5. Ibid., and Office of Technology Assessment, *Industry, Technology, and the Environment*, pp. 8–9.

6. Office of Technology Assessment, *Industry, Technology, and the Environment*.

7. Ibid., p. 9, and Gory Davis et al., "Car Recycling and Environmental Improvement in Western Europe," report prepared for Saturn Corporation, University of Tennessee Center for Clean Products and Clean Technologies, February 1993.

8. This case study comes from Strategic Environmental Associates, Somerville, MA, 1994.

9. Office of Technology Assessment, *Industry, Technology, and the Environment*, pp. 9–10, 318–20.

10. The quotations in this section come from Adam Kahane, "Global Scenarios for the Energy Industry: Challenge and Response," a selected

paper published in January 1991 by Shell's Group Public Affairs in London.

11. Ibid., p. 4.

12. Robert A. Frosch and Nicholas E. Gallopoulos, "Strategies for Manufacturing," *Scientific American,* September 1989, pp. 144–52. See also "Industrial Ecology Route to Slow Global Change Proposed," *Chemical and Engineering News,* August 24, 1992, pp. 7–14.

13. This discussion is based on Hardin Tibbs, "Industrial Ecology: An Environmental Agenda for Industry," *Whole Earth Review,* Winter 1992, pp. 39–40; see also Tibbs, "Industrial Ecology," *Pollution Prevention Review,* September 1992; Tibbs, *Innovation,* Special 1992 issue, pp. 37–40; and Tibbs, "Industrial Ecology" (Cambridge, MA: Arthur D. Little, 1991). The figure was provided by Hardin Tibbs.

14. Tibbs, "Industrial Ecology," pp. 14–15.

CONCLUSION: MANAGEMENT, THE ENVIRONMENT, AND JOBS

1. "With Cold War Over, Many Western Leaders Face Domestic Malaise," *Wall Street Journal,* July 3, 1992, p. A1.

2. Charles E. Teclaw, "Estimate of Returns on DOE Investments in the Industrial Waste Reduction Program," Los Alamos National Laboratory, Los Alamos, NM, May 1993, and *Industrial Waste Reduction Program Economic Evaluation,* CONSAD Research, January 1993.

3. U.S. Congress, Office of Technology Assessment, *Industry, Technology, and the Environment: Competitive Challenges and Business* Opportunities, OTA-ITE-58 (Washington, DC: U.S. Government Printing Office, January 1994), pp. 10, 319–20.

APPENDIX: YOU JUST DON'T UNDERSTAND: U.S. MISPERCEPTIONS OF JAPANESE SUCCESS

1. Amal Kumar Naj, "Some Manufacturers Drop Efforts to Adopt Japanese Techniques," *Wall Street Journal,* May 7, 1993, pp. A1, A6.

Bibliography

The following books are recommended for those who wish to apply systems thinking for redesigning processes to improve productivity and reduce pollution.

Covey, Stephen R. *The Seven Habits of Highly Effective People.* New York: Simon & Schuster, 1989.

Dorfman, Mark, et al. *Environmental Dividends: Cutting More Chemical Wastes.* New York: INFORM, 1992.

Ford, Henry. *Today and Tomorrow.* 1926. Reprint, Cambridge, MA: Productivity Press, 1988.

Hawken, Paul. *The Ecology of Commerce: A Declaration of Sustainability.* New York: Harper Business, 1993.

Hayes, Robert, Steven Wheelwright, and Kim Clark. *Dynamic Manufacturing: Creating the Learning Organization.* New York: Free Press, 1988.

Hirschhorn, Joel S., and Kirsten U. Oldenburg. *Prosperity Without Pollution: The Prevention Strategy for Industry and Consumers.* New York: Van Nostrand Reinhold, 1991.

Imai, Masaaki. *Kaizen: The Key to Japan's Competitive Success.* New York: McGraw-Hill, 1986.

Japan Human Relations Association. *The Idea Book: Improvement Through Total Employee Involvement.* Cambridge, MA: Productivity Press, 1988.

Kuhn, Thomas. *The Structure of Scientific Revolutions*, 2d edition. Chicago: University of Chicago Press, 1970.

Makower, Joel. *The E-Factor: The Bottom-line Approach to Environmentally Responsive Business*. New York: Times Books, 1993.

Martin, Larry. *Proven Profits from Pollution Prevention, Vol. 2*. Washington, DC: Institute for Local Self-reliance, 1989.

Meadows, Donella et al. *Beyond the Limits*. Post Mills, VT: Chelsea Green, 1992.

Ohno, Taiichi. *Toyota Production System: Beyond Large-Scale Production*. Cambridge, MA: Productivity Press, 1988.

Reich, Robert B. *The Work of Nations. Preparing Ourselves for 21st Century Capitalism*. New York: Alfred A. Knopf, 1991.

Romm, Joseph. *The Once and Future Superpower: How to Restore America's Economic, Energy, and Environmental Security*. New York: William Morrow, 1992.

Schmidheiny, Stephan. *Changing Course: A Global Business Perspective on Development and the Environment*. Cambridge, MA: MIT Press, 1992.

Schwartz, Peter. *The Art of the Long View*. New York: Doubleday, 1991.

Senge, Peter. *The Fifth Discipline: The Art and Practice of the Learning Organization*. New York: Doubleday, 1990.

Shingo, Shigeo. *Modern Approaches to Manufacturing Improvement: The Shingo System*. Ed. Alan Robinson. Cambridge, MA: Productivity Press, 1990.

————. *The Shingo Production Management System: Improving Process Functions*. Cambridge, MA: Productivity Press, 1992.

Smart, Bruce, ed. *Beyond Compliance: A New Industry View of the Environment*. Washington, DC: World Resources Institute, 1992.

Stalk, George, and Thomas Hout. *Competing Against Time: How Time-based Competition Is Reshaping Global Markets*. New York: Free Press, 1990.

U.S. Congress, Office of Technology Assessment. *Green Products by Design: Choices for a Cleaner Environment*. OTA-E-541. Washington, DC: U.S. Government Printing Office, October 1992.

————. *Industry, Technology, and the Environment: Competitive Challenges and Business Opportunities.* OTA-ITE-58. Washington, DC: U.S. Government Printing Office, January 1994.

Wann, David. *Biologic: Environmental Protection by Design.* Boulder, CO: Johnson Books, 1990.

Many individual articles, studies, and reports cited in the text are also useful, as are the journals *Pollution Prevention Review* and *Total Quality Environmental Management,* published by Executive Enterprises in New York.

Index

absenteeism, and office design, 92, 95, 98–99
Airbus Industries, 76
air conditioning, *see* heating, ventilation, and air conditioning (HVAC)
Aitken, Don, 98, 99
Allaire, Paul, 130, 145
Allenby, Braden, 138
American Express, 106
American Institute for Total Productive Maintenance, 158–59
anomalies, 59
Apple Computer, 84, 167
Arrow Automotive Industries, Inc., 144
Art of the Long View, The (Schwartz), 118–19
Art of War, The (Tzu), 109, 119
Artzt, Edwin L., 39–40
Asnaes, 173–75
assembly lines, moving, 19, 22
assumptions, challenging, 120–21
atriums, 65, 96
AT&T, 10, 45, 85, 119, 138
Avedissian, Jacques, 98

Baxter Healthcare Corporation, 5, 159–61
benchmarking, 52, 54, 76, 111–12, 122, 158–59
Beyond the Limits (Meadows), 119
biophilia, 103
Biophilia Hypothesis, The (Wilson), 103
Boeing, 4–5, 74–76, 119
Bonneville Power Administration, 37
Borsboom, Ton, 142
Boston Edison, 60
Boyd, John, 107, 109, 122, 125
Browning, Bill, 95
BSW Architects, 99
Buckeye Cellulose Corporation, 161–62
building managers, importance of, 68
Burt Hill Kosar Rittelman Associates, 87–88

Caldwell, Philip, 22
Canion, Rod, 69, 70
Canon, 53
Carter, John, 160

cascading, 61, 161–65
Cassens, Steve, 75
Cassidy, John, 185
cause-and-effect diagrams, 6–7, 57
ceilings, *xv–xvii*
Center for Resource Management, 99
change
  paradigms of, 13–14
  resistance to, 10–13
chlorofluorocarbons (CFCs), 6–7, 45, 61–62, 71, 160
Ciba-Geigy, 131
Clark, Kim, 26–27, 131–32
clean production, 6, 7, 16, 18, 130–66
  at Ford Motor Company, 18
  lean and clean design for, 137–41
  and life-cycle analysis, 33, 36
  and productivity, 29–30
  see also factory productivity
clean technology, 18
Clinton, Bill, 177
cogeneration systems, 95, 162–64
Colorado Pollution Prevention Partnership, 137, 147
communication
  with customers, 39–40, 46, 110–11, 143
  with employees, 40–41, 57, 112–13
Compaq Computer, 10, 31–33, 63–71
Competing Against Time (Stalk and Hout), 105–6
complexity, detail versus dynamic, 34–35
computers
  and energy efficiency, 83–85
  and productivity, 106
Comstock building (Pittsburgh), 5, 87–89, 127
Control Data, 83

cross-functional teams, 7, 48, 126–28, 180–81
  in building design, 88, 96
  compartmentalization versus, 126
  at Martin Marietta, 134
  tips for successful, 127–28
  see also teamwork
Crowley, Joseph, 60–62
customers, communication with, 39–40, 46, 110–11, 143
customer service, 11
Czege, Huba Wass de, 122

Dai Nippon, 155
daylighting, 79, 103
  benefits of, 40
  at Compaq Computer, 32, 65
  at Lockheed's Building 157, 96–97, 98
  at Nederlandsche Midden-standsbank (NMB Bank, Holland), 95
  at Wal-Mart's Eco-Mart, 99
decision making, 124–26
defect prevention
  correction versus, 12
  "end-of-pipe" approach to, 12–13
  and pollution prevention, 102
  rewarding employees for, 44
  see also quality
delegation, 125
Dell Computer, 70
demand-side management, 41
Deming, W. Edwards, 14, 22
Denmark, 173–75
design
  for the environment, *see* design for the environment
  lean and clean, principles of, 137–41
  for maintainability, 154–59
  office, *see* office design

process redesign, 36–42, 48, 137–42, 152, 181–82
product redesign, 139–41, 142, 148–54, 181–82
for remanufacture, 143–46
for resilience, 128–29
for simplicity, 141–43
design for the environment
and absenteeism, 92, 95, 98–99
at Compaq Computer, 10, 31–33, 70–71
cross-functional teams in, 7, 48, 88, 96, 126–28, 134, 180–81
and life-cycle analysis, 33
lighting in, *see* lighting
at Lockheed's Building 157, 4, 5, 93, 96–99
at Nederlandsche Midden-standsbank (NMB Bank, Holland), 95–96
office, 40, 83–85, 87–88, 90–104, 122
at Pennsylvania Power & Lighting Company, 90–92
and productivity, 29–30
at the Reno, Nevada post office, *xv–xviii*, 93
requirements of, 139
at Superior Die Set Corporation, 92–93
thermal-storage systems, 99–101
trial, 122
at UKettle, 5, 139–41
at Wal-Mart Eco-Mart, 99–100
at West Bend Mutual Insurance Company, 100–102
workstations in, 97, 98, 101–2
see also energy efficiency; pollution prevention; waste prevention
detail complexity, 34–35
DeVries, Doug, 82–83, 158
dissent, encouraging, 119–20

division of labor, 24–25
Dow Chemical, 4, 10, 39, 42–43, 119, 123, 177, 180
downsizing, 26
problems of, 8–9
trend toward, 7–8
DuPont, 10, 119
dynamic complexity, 34–35
Dynamic Manufacturing (Hayes, Wheelwright, and Clark), 24, 131–32, 141

Eastman Kodak, 53, 119
Ecology of Commerce, The (Hawken), 119
Elkhart General Hospital, 74
employees, 50–62
communication with, 40–41, 57, 112–13
contests for, 42–43, 123
control over environment, 101–2, 103
involvement of, systems approach to, 48–49, 58–59, 112–13
saving jobs of, 8–10, 51–53, 59–62, 177–82
suggestion systems, 9, 29, 51–58, 161
teamwork with, 33, 39, 40–43, 49, 50–59
training, 39, 44, 48, 57, 68, 122–24, 128, 156
wages of, 48–49, 51–53, 68, 177–78
see also jobs; productivity improvement; quality
employee suggestion systems, 51–58
at Baxter Healthcare, 161
at Republic Engineered Steels, 9
at Toyota, 29
end-user focus, 12–13, 39–41, 143

end-user focus (*cont'd*)
  in energy efficiency, 77, 79–83
  and lighting, 73–83, 99, 101
Energy, U.S. Department of, 179
energy conservation
  energy efficiency versus, 11,
    73
  and natural resource availability,
    11–12
energy efficiency, 72–89, 148–54,
    168
  and building managers, 68
  cogeneration, 95, 162–64
  at Compaq Computer, 66–71
  components of, 6
  concept of, 72–73
  conservation versus, 11, 73
  at Dow Chemical, 42–43
  end-user focus in, 77, 79–83
  in heating, ventilation, and air-
    conditioning (HVAC) sys-
    tems, 6, 65–67, 68, 85–86,
    87, 99–101
  and life-cycle analysis, 36
  and lighting, 73–83, 99, 101
  and motor system redesign,
    148–52
  and office equipment, 83–85
  and productivity improvement,
    61–62
  at the Reno, Nevada post office,
    xv–xviii, 93
  saving jobs through, 9, 59–62,
    177–82
  systems approach to, 6, 87–89
  utility company role in, 37, 60,
    76–78, 82, 90–92, 101
  and waste prevention, 41
Energy Star Computer program
    (EPA), 70, 84
environment, as term, 33–34
environmental benchmarking, 112
environmental design, *see* design
  for the environment

Environmentally   Responsive
  Workstations (ERWs), 101–2
Environmental Protection Agency
  (EPA)
  Energy Star Computer program,
    70, 84
  "Green Lights" program, 73–74,
    78, 160
  and hazardous waste, 133, 159
environmental reengineering, in
  process redesign, 37, 42
*Environmental Self-Assessment Pro-
  gram* (GEMI), 165–66

factory productivity, 130–66
  assessment of, 165–66
  cascading in, 61, 161–65
  designing for remanufacture in,
    143–46
  designing for simplicity in,
    141–43
  energy efficiency in, 148–54
  and fast-cycle approach, 105–9
  and Hawthorne effect, *xvii–xviii*,
    93–94
  interchangeable parts in, 19, 22
  life-cycle analysis in, 146–48
  moving assembly lines in, 19, 22
  paradigm shift for, 132–37,
    165–66
  preventive   maintenance   in,
    154–59
  and process redesign, 36–42, 48,
    137–41, 142, 152, 181–82
  and product redesign, 139–41,
    142, 148–54, 181–82
  recycling in, 18, 51, 130, 131,
    132, 159–61
  see also clean production; pollu-
    tion prevention; waste preven-
    tion
fast-cycle approach, 36, 69
  and cross-functional teams, 7,
    48, 126–28, 134

designing for resilience in, 128–29
importance of, 107–9
and learning process, 140
to manufacturing, 105–9
and pollution prevention, 128
to warfare, 107, 109, 122, 125
see also life-cycle analysis; product cycle time
Federal-Mogul, 183–84
feedback, 165
and cross-functional teams, 127
customer, 39–40, 46, 110–11, 143
employee, 40–41, 57, 112–13
and preventive maintenance, 157
in suggestion systems, 9, 29, 51–58, 161
in systems approach, 43–49
Feigenbaum, Armand, 22–23
Fifth Discipline, The (Senge), 34, 119
Fisher Scientific, 48, 127, 131
Fitch Richardson Smith, 139–41, 142–43
Flanigan, Ted, 64
flexible manufacturing, 110, 123
Ford, Henry, 27, 51, 182
and lean and clean management, 16, 17–22
on salvage, 130, 162–63
Ford Motor Company
early years, 16, 17–22
Team Taurus, 127–28
Forsyth, Fred, 84
Friedman, Lawrence, 74–76
Fritz, Robert, 110

Garvin, David, 57
General Electric, 119, 184–85
General Motors, 27, 57, 106, 113–16, 173
GE Plastics, 139, 145–46

Germany, 139, 169
Geus, Arie De, 124
Gilbreth, Frank B., 23, 25–26, 28–29
Gilbreth, Lillian, 23
Global Business Network, 118–19, 173
Global Environmental Management Initiative, 119, 166
Global Turnkey Systems (GTS), 85–86, 164–65
GoldenBear Cogen Inc., 163–64
Gonsalves, Ron, 37
Gray, Alfred, 119
Green Aid plan (Japan), 168–69
green design, *see* design for the environment
Green Gold (Moore and Miller), 167
"Green Lights" program (EPA), 73–74, 78, 160
Green Products by Design (Office of Technology Assessment), 138–39
Gribi, John, 31, 64

Harrah's Hotel and Casino, 4, 5, 86, 120–21
Hawken, Paul, 119
Haworth, Inc., 159
Hawthorne effect, *xvii–xviii*
misinterpretation of, 93–94
Hayes, Robert, 26–27, 131–32
hazardous waste, 133, 159
heating, ventilation, and air conditioning (HVAC), 6, 68, 85–86, 87
at Compaq Computer, 65–67
thermal-generation systems, 99–101
heat pumps, 105–7
Heede, Rick, 84–85
Henkel, 169
Hewlett Packard, 43, 84, 147

Home Depot, 121–22, 146
Honda, Soichiro, 50
Honda Motors, 123, 128
Hussein, Saddam, 109
HVAC, *see* heating, ventilation, and air conditioning (HVAC)
Hyde, Isaac P., 83
Hyde Tools, 82–83, 158

Iansiti, Marco, 127
IBM, 63–64, 69, 85, 106
Idea Book, The (Japan Human Relations Association), 54
Imai, Masaaki, 41, 156
industrial ecology, 34, 172–75
Industrial Revolution, 24, 25
Industry, Technology, and the Environment (Office of Technology Assessment), 12–13, 167–68
INFORM, 130–31
Inland Rome Lumber Company, 151
Institute for Local Self-reliance, 161–62
interchangeable parts, 19, 22
interconnections, 126
and fast-cycle approach, 129
in systems approach, 33–34
Introduction to TPM (Nakajima), 159
IRT Environment, 64

Japan
and flexible manufacturing, 110, 123
Green Aid plan, 168–69
and lean production, 16, 17, 36
and quality, 22–23
suggestion systems in, 53–54, 57–58
systems approach in, 16, 17, 22–29
Japan Human Relations Association, 53, 54

Japan Institute of Plant Maintenance, 154
jobs
creating, 177–82
and downsizing trend, 7–8
and energy efficiency, 59–62
in lean and clean management, 8–10, 51–53, 59–62, 177–82
loss of, preventing, 8–10, 51–53, 59–62, 177–82
and wages, 48–49, 51–53, 68, 177–78
see also employees
Juran, Joseph, 22–23
just-in-time production system, 27, 36, 41, 108

Kahane, A., 171
kaizen, 41–42, 53
Kaizen (Imai), 156
Katz, Harry, 123
Kay, Alan, 167
Kelly, Harold, 51
Kennedy, John F., 113
Kimura, Ben, 98
Kiuchi, Tachi, 106–7
Korean War, 107, 109
Kraft General Foods, 59–62
Kuhn, Thomas
on anomalies, 59
and paradigm shifts, 13–14

labor, *see* employees
landscaping, 99
laptop computers, 84–85
Larsen, George, 128, 133
Lauret, Ronald W., 102
layoffs, *see* downsizing
lean and clean design, and process redesign, 137–41
lean and clean management, *xviii,* 3–4, 7, 16, 17–22
barriers to, 12–13

and environmentally proactive companies, 4–6
saving jobs with, 8–10, 51–53, 59–62, 177–82
lean production, 6, 7, 16, 17, 36, 41–42
least-cost approach, 40
Lee, Eng Lock, 65–67
Leo J. Daly, *xv*, 96–99
Liebe, Tie, 95
life-cycle analysis
  in building design, 87–88, 90–94
  at Compaq Computer, 64, 65–66
  of employees, 48–49
  in factory productivity, 146–48
  in pollution prevention, 33, 146–47
  in systems analysis, 31–33, 36, 47–48, 121–22
  and time, 31–33, 36
  and total productive maintenance, 158
  and Total Quality Management (TQM), 14
lighting, *xv–xvii*, 73–83
  at Compaq Computer, 32, 65, 69
  daylighting, 32, 40, 65, 79, 95–99, 103
  end-user focus and, 77, 79–83
  and energy efficiency, 73–83, 99, 101
  EPA "Green Lights" program, 73–74, 78, 160
  hardware upgrades, 78–79
  payback analysis for, 79, 81, 92, 93
  at Pennsylvania Power & Lighting Company, 90–92
  and productivity improvement, *xv–xvii*, 6, 32, 73, 91–93, 97, 101–3
  and quality improvement, *xv–*

xvii, 74–76, 80, 82–83, 92, 102–3
  systems approach to, 6
  utility company role in, 60, 76–78, 82
  and waste prevention, 44, 46–47
Lockheed Building 157, 4, 5, 93, 96–99
Lovins, Amory, 40, 72, 79, 84–85, 119
LTV Corporation, 51

Machine That Changed the World, The (Womack et al.), 115
MacKenzie, Jim, 145
Magliozzi, Tom and Ray, 23
Maier, Russell, 56
maintenance, preventive, 154–59
maintenance prevention, 155
Martin Marietta, Astronautics Group, 10, 128, 132–37, 147
Massachusetts Institute of Technology, 41–42, 115
Matsushita, 53–54, 143
McDonough, William, 88, 99
McLean, Robert, *xv, xvii*
McMullen, Bob, 137
McPherson, Rene, 155
Meadows, Donella, 36, 119
  on feedback, 45
  on systems, 33
Merck, 119
Microsoft, 146
Milini, Ray, 52
Miller, Alan, 167
Mitroff, Ian, 113–14
Mitsubishi Electric Company, 105–7
Model-A Ford, 22
Model-T Ford, 17, 21–22, 23
modularity, 145–46
Moore, Curtis, 167
Morabito, Dick, 145
Morehouse, E. Thomas, Jr., 146

motion sensors, 101
Motorola, 112, 119–20, 123–24, 129
motor systems, redesign of, 148–52
moving assembly lines, 19, 22
Musone, Fred, 184
Musser, Rob, 135

Nakajima, Seiichi, 159
National Aeronautics and Space Administration (NASA), 113, 132
Navarini, Joanne, 98
Neal, Richard, 109
Nederlandsche Middenstandsbank (NMB Bank, Holland), 95–96
Nelson, Kenneth, 42
Netherlands, 169–70, 181
Nissan, 155

observation, 110–13
office design, 40, 90–104
    and absenteeism, 92, 95, 98–99
    Comstock building (Pittsburgh), 5, 87–89, 127
    and energy efficiency of equipment, 83–85
    Hawthorne effect, *xvii–xviii*, 93–94
    life-cycle approach to, 87–88, 90–94
    at Lockheed's Building 157, 4, 5, 93, 96–99
    at Nederlandsche Middenstandsbank (NMB Bank, Holland), 95–96
    at Pennsylvania Power & Lighting Company, 90–92
    at Superior Die Set Corporation, 92–93
    at West Bend Mutual Insurance Company, 100–102
    workstations in, 97, 98, 101–2

see also lighting; heating, ventilation, and air conditioning (HVAC)
Office of Technology Assessment (OTA), 12–13, 76, 138–39, 168, 169
Ohno, Taiichi, 20, 23, 24, 27–29, 93, 151
Once and Future Superpower, The (Romm), 167
O-O-D-A loop, 109–29
operations, process versus, 25–27
orientation, 113–24
    and Royal Dutch/Shell, 116–18

Page, Thomas, 78
paper, cascading, 164–65
paradigms, 113–24
    of scientific change, 13–14
    shifts in, 13–14, 69–70, 132–37, 165–66
    of time waste, 19–21, 22
    of waste and natural resources, 11–12
    worldview of managers, 13–15, 60, 118–22
Pareto charts, 57
Path of Least Resistance, The (Fritz), 110
payback analysis
    for cogeneration projects, 164
    at Compaq Computer, 31–33, 66
    at Dow Chemical, 43
    for lighting improvements, 79, 81, 92, 93
    in motor redesign, 149, 152
    for pollution prevention, 4
    by Regal Fruit Co-op, 37
    and Reno, Nevada post office, *xvii*
    at Sealtest, 60
Pennsylvania Power & Lighting Company, 90–94

Perkins, Ron, 31–33, 63, 64–67, 69, 150
Persian Gulf War, 109, 122
Petsch, Gregory, 70
Pfeiffer, Eckhard, 70
pinch technology, 151–52
pollution prevention, 109
  benefits of, 177–82
  and clean production, 6
  and defect prevention, 102
  and energy efficiency, 73–74
  and environmentally proactive companies, 4–6
  and fast-cycle approach, 128
  at Fisher Scientific, 48
  lack of focus on, 12–13
  and life-cycle analysis, 33, 146–47
  and Martin Marietta Astronautics Group, 132–37
  payback analysis for, 4
  and productivity improvement, 36–42, 130–32
  rate of return on, 9–10
  and Reno, Nevada post office, *xvii, xviii*
  systematic process design for, 10–15
  systems approach to, 6–7, 132–37
  trend toward, 167–70
  and waste prevention, 3–4, 44, 45
pollution prevention circles, 125–26
Polymer Solutions, 139–41
Porter, Michael, 3
predictive maintenance, 155
prevention
  at Compaq Computer, 63–71
  focus on, 38
  and proactive approach, 38–39, 43–44
  training workers in, 39

see also pollution prevention; waste prevention
prevention engineering, 143
preventive maintenance, 154–59
problem solving skills, 56–57
process, operations versus, 25–27
process design, and systems approach, 48
process redesign, 142, 181–82
  lean and clean design in, 137–41
  at Regal Fruit Co-op, 36–37, 152
  and systems approach, 36–42, 48
Procter & Gamble, 39–40, 119
product cycle time, 36, 46
  in the Comstock building, 5, 87–89, 127
  and cross-functional teams, 7, 48, 88, 126–28, 134
  and flexible manufacturing, 110, 123
  and the Mitsubishi heat pump, 105–7
  O-O-D-A loop in, 109–29
  and scenario planning, 116–19
  in warfare, 107, 109, 122, 125
  see also defect prevention; fast-cycle approach; life-cycle analysis
productivity improvement
  and building design, 101–2, 103
  and clean production, 29–30
  at Compaq Computer, 32–33, 65
  employee contributions to, 39, 40–43, 50–59
  and energy efficiency, 61–62
  and Hawthorne effect, *xvii–xviii*, 93–94
  and lighting, *xv–xvii*, 6, 32, 73, 91–93, 97, 101–3
  at Pennsylvania Power & Lighting Company, 90–94

productivity improvement (*cont'd*)
　and pollution prevention, 36–42,
　130–32
　at the Reno, Nevada post office,
　xv–xviii, 93
　systems approach to, 11
　see also factory productivity; office design
product redesign, 139–41, 142,
　148–54, 181–82

quality
　"end-of-pipe" approach to,
　12–13
　and improved lighting, *xv–xvii*,
　74–76, 80, 82–83, 92, 102–3
　and systems approach, 22–23,
　33–36, 38–42, 110–11
　see also Total Quality Management (TQM)
quality circles, 22–23, 64, 125–26,
　184–85

rate of return analysis, 9–10, 39
recycling, 51, 159–61
　and environmentally proactive
　companies, 4–6
　in factory productivity, 18, 51,
　130, 131, 132, 159–61
redesign
　process, 36–42, 48, 137–42,
　152, 181–82
　product, 139–41, 142, 148–54,
　181–82
reengineering, 37, 42
Regal Fruit Co-op, 36–37, 152
Reich, Robert, 180
　on downsizing, 8
　on symbolic analysis, 31
　on systems analysis, 33
remanufacture, design for, 143–46
renewable energy, 168
Reno, Nevada post office, *xv–xviii*,
　93

Republic Engineered Steels, 4, 5,
　9, 51–53, 54, 56, 59, 178
resilience, designing for, 128–29
return on investment analysis,
　9–10, 39
Robinson, Russell, 97
Rocky Mountain Institute, 66, 84–
　85, 95, 99, 119
Royal Dutch/Shell, 116–18, 124,
　170–72
Russell, Allen, 90–92

St. Francis Hospital, 74
salvage, 130, 162–63
　see also recycling
San Diego Gas & Electric
　(SDG&E), 76–78
Saturn Corporation, 57, 169
Scanlon, Paul, 87–88
scenario planning, 116–19,
　170–72
School of Advanced Military Studies (SAMS), 122
Schwartz, Peter, 118–19
Schwarzkopf, Norman, 122
Sealtest, 9, 59–62, 177–78
Sears, 106
Seay, Tom, 100
Senge, Peter, 34, 35, 113, 119
Shell International, 116–18, 124,
　170–72
Shingo, Shigeo, 23–27, 29, 54, 93,
　128–29
simplicity, designing for, 141–43
Sinnreich, Richard, 122
Small Business Administration
　(SBA), 180
Smith, Adam, 24
Soft Energy Paths (Lovins), 72
Southwire Corporation, 9, 152–54,
　178
specialization, 24–25
　at Ford Motor Company, 19, 21
Spender, J. A., 20–21

Square D Company, 142–43
Steinbacher, Herbert, 159
Steinbacher, Norma, 159
Steuben Foods, Inc., 151–52
Stimac, Gary, 69
streamlining, 26
Structure of Scientific Revolutions, The (Kuhn), 13–14
suggestion systems, 9, 29, 51–58, 161
  benefits of, 57–58
  establishing, 54–58
  feedback in, 9, 29, 51–58, 161
  in Japan, 53–54
Superior Die Set Corporation, 92–93
Supersymmetry USA, Inc., 64, 65–67, 150
systematic process improvement, 6
systems approach, 6–7, 31–49, 105–29
  barriers to use of, 34–35
  and change, resistance to, 10–13
  continuous improvement in, 41–43
  to employee involvement, 48–49, 58–59, 112–13
  to energy efficiency, 6, 87–89
  fast-cycle approach in, 105–9
  feedback in, 43–49
  focus on end results in, 39–41
  importance of, 48–49
  Japanese, 16, 17, 22–29
  keys to, 38–42
  life-cycle analysis in, 31–33, 36, 47–48, 121–22
  and O-O-D-A loop, 109–29
  to pollution prevention, 6–7, 132–37
  as proactive, 38–39, 43–44
  in process redesign, 36–42, 48
  to productivity improvement, 11
  to quality, 22–23, 33–36, 38–42, 110–11
  systems thinking in, 33–36, 38–42

Tanii, Akio, 143
tax code, 181
Taylor, Frederick, 23, 24
teamwork, 33
  with employees, 33, 39, 40–43, 49, 50–59
  and suggestion systems, 54
  see also cross-functional teams
Teeters, Ken, 86, 120–21
telecommuting, 85
Tenneco, 119
thermal-generation systems, 99–101
3M, 10
Tibbs, Hardin, 173
time
  life-cycle analysis and, 31–33, 36
  in systems approach, 34–36
  and Total Quality Management (TQM), 35–36
  and waste prevention, 6, 17, 19–20
time-study techniques, 23, 24
Today and Tomorrow (Ford), 17–18
Topy Industries, Ayase Works, 156–57, 158
total factor productivity (TFP), 131–32
total productive maintenance (TPM), 154–59
Total Quality Environmental Management, *see* lean and clean management
Total Quality Management (TQM), 6–7, 46, 65, 137
  at Boeing, 75–76
  customer contact in, 110–11
  failure of, reasons for, 13–15, 29–30, 35–36
  and layoffs, 8–9

Total Quality Management (*cont'd*)
  measurement in, 112
  at Sealtest, 60–61
Toyoda, Eiji, 22
Toyota, 20, 22, 23, 27–29, 128–29
TPM for America (Steinbacher and
  Steinbacher), 159
training, employee, 39, 44, 48, 57,
  68, 122–24, 128, 156
Tzu, Sun, 70, 109, 119

UKettle, 5, 139–41
Ulrich, Roger, 103
Union Carbide, 119
United Technologies, 185
U.S. Total Employee Involvement
  Institute, 54
utility companies
  and energy-efficiency programs,
    37, 60, 76–78, 82, 90–92, 101
  peak power demands, 151, 153

Wack, Pierre, 116–19
wages, 48–49, 51–53, 68, 177–78
Wal-Mart, 99–100
war, art of, 70, 107, 109, 119, 122,
  125
Warner, Aisin, 54
Wascher, Uwe, 139
waste prevention
  at Baxter Healthcare Corpora-
    tion, 5, 159–61
  at Dow Chemical, 42–43
  employee role in, 51
  and energy efficiency, 41
  at Ford Motor Company, 18–20
  and lighting, 44, 46–47
  and pollution prevention, 3–4,
    36–42, 44, 45, 130–32

and productivity improvement,
  130–32
in systems approach, 41
and time, 6, 17, 19–20
at Toyota, 27–28
trend toward, 167–70
water, cascading, 161–62
water-efficiency programs, 51–52
Watkins, James, 60
Wealth of Nations, The (Smith),
  24
Weld, William, 60
West Bend Mutual Insurance Company, 100–102
Western Electric, Hawthorne studies, *xvii–xviii*, 93–94
Wheelwright, Steven, 131–32
Whirlpool, 142
white noise, 101
why, asking five times, 28–29, 57
Wiggenhorn, William, 123
Wilson, Edward O., 103
Windheim, Lee, *xv–xvi*, 93, 96–97,
  98–99
windows, 65, 85, 95, 96
Wohlstetter, Roberta, 116
workers, *see* employees
Work of Nations, The (Reich), 31,
  180
workstations, in office design, 97,
  98, 101–2
World Bank, 168
Worldwide   Office   Environment
  Index Survey, 73
Wright, Frank Lloyd, 90, 93

Xerox, 10, 40, 130, 144–45, 145

Yakmalian, Karney, 119, 124